ALBUM OF SCIENCE

The Physical Sciences
in the
Twentieth Century

ALBUM

OF

SCIENCE

The Physical Sciences in the Twentieth Century

OWEN GINGERICH

I. B. Cohen, General Editor, Albums of Science

CHARLES SCRIBNER'S SONS

New York

1. (*Frontispiece*) Earth rising above the lunar horizon. The twentieth century brought new ways of looking at the physical world. Theorists began speaking of relativity and the uncertainty principle, particles within particles within particles, black holes, and continents adrift. Advances in technology yielded astonishingly powerful research instruments, while computers extended the mind's grasp, and rockets made possible scientific observation beyond the earth's confines. Here is the lunar module ascent stage of the 1969 U.S. Apollo 11 mission as it moves up to dock with the orbiting command service module, from which this photograph was taken. Man has set foot on the moon, and *Eagle* is coming back toward its triumphant conclusion.

Copyright © 1989 Macmillan Publishing Company

Library of Congress Cataloging-in-Publication Data
Gingerich, Owen.
   The physical sciences in the twentieth century.

   (Album of science)
   Bibliography: p.
   Includes index.
   1. Physical sciences—History—20th century.
I. Title. II. Series.
Q125.G495   1989   500.2'09'04   88-24007
ISBN 0-684-15497-8

Other volumes in the *Album of Science* series:
John E. Murdoch, *Antiquity and the Middle Ages*, ISBN 0-684-15496-X
I. Bernard Cohen, *From Leonardo to Lavoisier, 1450–1800*, ISBN 0-684-15377-7
L. Pearce Williams, *The Nineteenth Century*, ISBN 0-684-15047-6
Merriley Borell, *The Biological Sciences in the Twentieth Century*, ISBN 0-684-16483-3

Album of Science, five-volume set, ISBN 0-684-19074-5

Published simultaneously in Canada
by Collier Macmillan Canada, Inc.
Copyright under the Berne Convention

3 5 7 9 11 13 15 17 19 Q/C 20 18 16 14 12 10 8 6 4 2

Printed in the United States of America

Picture research by John Schultz—PAR/NYC

To David Latham

and a generation of
teaching fellows
in our course
"The Astronomical Perspective"

Contents

Acknowledgments

The task of selecting over 400 pictures and producing relevant and accurate captions to represent twentieth-century physical science proved more daunting than I ever anticipated, so I am particularly grateful to those colleagues who joined in on the writing of chapter introductions when the task seemed overwhelming. Their names are mentioned in the list of contributors; in many instances they helped check out the captions, and they suggested some of the items in the Guide to Further Reading. Stephen Brush, especially, took on a substantial share of the introductory essays, and his generous assistance enabled us to resolve several seeming contradictions in the picture captions.

I. Bernard Cohen originally conceived of this project and offered many helpful suggestions concerning its organization, as well as several ingenious pictorial ideas. It was he who proposed that the work might move more expeditiously if colleagues took on some of the writing.

Among others who contributed ideas and suggestions were Philip Morrison, Victor Weisskopf, Leonard Nash, Ursula Marvin, Frederick Nebeker, David Latham, Woodruff T. Sullivan III, and C. Stewart Gillmor. More people than I can even remember helped out in obtaining specific pictures or specific information required to make the pictures meaningful. These included Spencer Weart (American Institute of Physics), Richard Dreiser (Yerkes Observatory Photographic Services), many colleagues at the Harvard-Smithsonian Center for Astrophysics, and the reference staff of Harvard's Cabot Science Library. The Nobel Laureates section is based on *Collier's Encyclopedia*.

One of the most exciting moments of the project came when I first saw the picture layouts, and for the excellent design work I thank especially Marvin Friedman; for other design and production details I gratefully acknowledge the rest of the team at Macmillan as well: Zelda Haber, Joan Gampert, Lee Goldstein, Mynette Green, Valerie James, Marc Sferrazza, and Adrienne Weiss. John Schultz undertook the major work of obtaining the pictures, and others were procured by Macmillan photo editor Joyce Deyo.

Despite all this excellent assistance, the project surely would have foundered except for the continual attention and efficient organization brought by Richard Hantula, the Macmillan editor who worked with me very closely throughout this entire endeavor. He always knew how to find the relevant picture to fill a gap, where one sentence contradicted another, or when we failed to make a scientific point clear enough for the general reader. He watched over the designers when they turned figures upside down, over the contributors when they forgot to identify the people they cited, and over me when I tried to mention in the text pictures that had fallen by the wayside for lack of space. With feigned despondency he cajoled one piece after another of this production from me, and now in its successful conclusion, he deserves immense gratitude from his employer, from me, and from readers who enjoy the book. I thank him heartily for making the book a reality.

Contributors of Chapter Introductions

William Aspray, Associate Director, Charles Babbage Institute, University of Minnesota at Minneapolis; coeditor of *Papers of John von Neumann on Computers and Computing Theory* (1986) and *History and Philosophy of Modern Mathematics* (1988); editor of *Computing Before Computers* (1989). THE COMPUTER REVOLUTION.

Stephen G. Brush, Professor, Department of History and the Institute for Physical Science and Technology, University of Maryland; works include *Statistical Physics and the Atomic Theory of Matter from Boyle and Newton to Landau and Onsager* (1983) and *The History of Modern Science: A Guide to the Second Scientific Revolution, 1800–1950* (1988). RADIOACTIVITY AND THE NUCLEUS; ATOMIC CHEMISTRY; RELATIVITY; QUANTUM MECHANICS; THE AGE OF THE EARTH; CONTINENTAL DRIFT; NUCLEAR PHYSICS; NUCLEAR FISSION AND NUCLEAR FUSION; SCIENTISTS UNITE!

P. Thomas Carroll, Assistant Professor of History, Department of Science and Technology Studies, Rensselaer Polytechnic Institute; works include *Chemistry in America, 1876–1976: Historical Indicators* (coauthor; 1985). HOW ATOMS UNITE.

David DeVorkin, Curator, History of Astronomy, National Air and Space Museum; works include *Modern Astronomy and Astrophysics: A Selected and Annotated Bibliographical Guide* (1982), *Practical Astronomy: Lectures on Time, Place, and Space* (1986), and articles on the history of astronomy and space science. MEN ON THE MOON; THE EXPLORATION OF SPACE; COMMUNICATING SCIENCE.

Robert Friedel, Associate Professor, Department of History, University of Maryland; books include *Pioneer Plastic: The Making and Selling of Celluloid* (1983) and *Edison's Electric Light: Biography of an Invention* (coauthor; 1986). CHEMICAL TECHNOLOGY; ELECTRONICS.

Simon Mitton, Editorial Director (Science, Technical, and Medical Publishing), Cambridge University Press and Fellow of St. Edmund's College, Cambridge; has authored and edited several books, including *The Cambridge Encyclopedia of Astronomy* (1977). RADAR AND RADIO ASTRONOMY.

Susan Schlee, science writer; books include *The Edge of an Unfamiliar World: A History of Oceanography* (1973) and *On Almost Any Wind: The Saga of the Oceanographic Research Vessel* Atlantis (1978). OCEANOGRAPHY.

Robert Seidel, Museum Administrator, Bradbury Science Museum, Los Alamos National Laboratory; Research Historian, Laser History Project; works include *Lawrence and His Laboratory: Nuclear Science at Berkeley* (1981). INSIDE THE NUCLEUS.

*F*oreword

The *Album of Science* series was conceived as a pictorial record of the growth of the scientific enterprise. By means of images it seeks both to portray what science has been like and to convey a sense of how science has been perceived by scientists and nonscientists in different periods. The most important advances in the exact or physical sciences of the twentieth century are represented in the book at hand, prepared by Owen Gingerich with the help of colleagues who contributed some of the chapter introductions. Like the four previous volumes in the series, however, the present book is not a mere record of great discoveries. A complete pictorial record of the "great discoveries" could not be produced within the *Album of Science* framework, since many important discoveries are of an abstract nature and are not easily portrayed through pictures.

Owen Gingerich's volume, like Merriley Borell's on the biological sciences in the twentieth century, differs in a significant way from the first three books in the series—*Antiquity and the Middle Ages* (by John E. Murdoch), *From Leonardo to Lavoisier* (by myself), and *The Nineteenth Century* (by L. Pearce Williams). This distinctive feature is that the whole time span covered has been recorded by photography. Consequently, most of the illustrations in the volume are photographs.

This is not to say that Owen Gingerich has refrained from drawing on other types of illustrative material. There are reproductions of telling pages from scientists' notes, as well as of segments of major printed works, including Einstein on relativity and Staudinger on polymerization, along with striking diagrams and drawings, such as Kamerlingh Onnes's schematic for his liquefier. There are reproductions of documents reflecting scientists's personal lives or the public's view of science. Helping to convey a sense of society's perception of science are reproductions of postcards, postage stamps, and works of art based on scientific theory or influenced by it in one way or another. The reader is not allowed to forget the wit manifested by those involved in the scientific endeavor: offered for examination is the first computer "bug," and a bumper sticker based on Heisenberg's uncertainty principle.

The photographs reproduced here differ in a major respect from the illustrations in all the other volumes, including the one on the twentieth-century biological sciences. This arises from the fact that throughout the whole of our century photography has not only provided a visual record of events in the development of scientific thought and experiment, but has also been a fundamental part of experiments themselves. For instance, the early study of X rays used photographic materials to record the data of experiments. Radioactivity was discovered through the action of a radioactive mineral on a photographic plate that had never been exposed to light. The first X-ray photograph ever made is reproduced in this volume, as is the original plate recording what came to be known as radioactivity. Additionally, photographic plates were used to record cosmic rays—events on the scale of atomic physics—as well as the occurrence of solar prominences, phenomena falling under the purview of astronomy. For the latter we have in the present volume both original photographs made early in the century by colleagues of George Ellery Hale and more recent ones made by Skylab.

Since the goal of the *Album of Science* is to show what science was like, rather than to provide a picture gallery of famous men and women, the book has few formal portraits. But the reader does see scientists at work in their laboratories or gathered in groups. As its companion volume did for the biological sciences, Owen Gingerich's book seeks to convey a sense of the whole environment of the physical sciences in our times. Both in astronomy and in physics the twentieth century has seen the development of large installations, requiring the services of many research scientists and technicians. The earliest large-scale, expensive centers of research were huge observatories, featuring ever bigger and more powerful telescopes. In more recent times the record for huge expenses has been held by high-energy physics. Furthermore, the use of satellites and spaceships has produced a new range of data of significance.

Owen Gingerich's book uses images from a wide variety of archival sources, and many of the pictures have not previously been reproduced in book form. As a group, the images give the viewer a sense of having been present when great discoveries were in the process of being made. But even the most striking photographs portraying the growth of the scientific enterprise cannot stand by themselves. The text prepared by Owen Gingerich and his colleagues supplies the necessary interpretation, delineating major concepts and events and noting their scientific and social significance.

The reader is constantly reminded that scientific research and discovery take place in, and have implications for, society at large. The volume displays the international character of scientific research and shows the growth of "Big Science." The author makes us aware of the significant endeavor to communicate scientific findings to fellow scientists and to the public at large.

A particularly notable feature of the present volume is that Owen Gingerich has depicted the close relations between the advancing physical sciences and new technology. In some instances, it is even difficult to make an easy distinction between the two. This characteristic of the physical sciences in the twentieth century may be most easily discerned in the chapter dealing with the development of the atom bomb and the rise of nuclear energy. Again and again in our century we have witnessed the development of new technologies or the improvement of old technologies, coming in a sudden and unanticipated fashion from fundamental investigations into nature. At the same time, as is made manifest in this book, many practical or technological problems have required fundamental research to provide the knowledge base for their solution.

The present volume on the physical sciences also shows an interesting paradox in the development of Big Science: Science requires the expenditure of large sums of money, primarily provided by government, in order to explore a smaller and smaller scale of nature. Penetration into microphysics has required macro-expenditure of funds on a scale wholly unprecedented.

All readers are fully aware of the dual character of the physical sciences in the twentieth century. As is displayed in the pages of this book, the physical sciences have brought us a deeper knowledge of the structure and forces of matter and have furnished us with the power to improve the quality of life, but they have also brought the power to destroy life itself. While discoveries in the realm of the atom have made possible a new source of energy, they have also produced the most terrible instrument of destruction the world has ever known. Moreover, even using the new source of energy raises questions of contamination and pollution. Thus, the quality of life is affected by decisions made about technology that are based on science.

Owen Gingerich's volume cannot resolve for us such questions, which impinge on the lives of all the earth's inhabitants. But the book can help us come to grips with them by increasing our understanding of the nature of the scientific process. Every reader will derive a deepened appreciation of the intellectual adventure and enterprise constituting the physical sciences in our century.

I. Bernard Cohen

Introduction

This *Album of Science* attempts to cram within its three hundred or so pages a visual survey of nearly ninety years of the physical sciences. Some of the photographs reproduced here were the very basis of scientific discoveries. Others show the equipment, the laboratories, the people involved. A few depict notebooks recording essential data or papers containing seminal findings. Some reveal the attitudes of the public to scientific advances or the response of artists or cartoonists. Altogether they document the flavor and moods of twentieth-century physical science.

In many ways the *Album* is like a Victorian scrapbook: here are two tickets to the opera, there lies the napkin from the ball, and on the next page, a snapshot of Uncle Henry by the waterfall. The scrapbook is erratic and opportunistic, with splendid mementos and curious omissions. Alas, no one thought to take a snapshot of Albert Einstein describing the universe as seen from a photon of light or of Robert Woodward with a model of the quinine molecule that he had just synthesized. The *Album* has no picture of the overthrow of parity conservation, no photograph of a black hole or a cosmic string, no view of the El Niño current of the Pacific Ocean. But here is a candid shot of Werner Heisenberg at breakfast with Niels Bohr (explaining the uncertainty principle?), and a breathtakingly memorable photograph from the first manned exploration of the moon. There, the evidence that gravity bends light, and on another page, the explosion at Trinity that changed the world. Here is a photo of a single atom, and there one of the largest known astronomical objects, a giant radio jet spurting from an active galaxy.

In the capricious archive of illustrations that history has left us, astronomy comes out best. It is a naturally pictorial subject, from early photographs of solar prominences and spiral nebulae to modern televised scenes from the strange worlds of Io and Miranda, to the wondrous electronic reconstructions of radio jets and intergalactic winds. The telescopes and spacecraft that have made these views possible are likewise photogenic, and they appear on the *Album*'s pages. But how can the immense strides in understanding the evolution of stars and their internal constitutions be depicted? Curiously enough, we come perhaps closest to the inside of a star in a scene from inside the earth, deep under Lake Erie, where scientists have detected neutrinos from the interior of a supernova.

Picture albums are like that: they represent hardware better than theoretical structures. The most brilliant physics of this century falls into the realm of abstract ideas: quantum mechanics, relativity, unified field theories. And these are not so easy to illustrate, particularly since by deliberate design the *Album* generally avoids portraits as such. Nevertheless, physicists engaged in energetic discussions—Rutherford, Einstein, Bohr, Heisenberg, Fermi, Feynman, to name a few—find appropriate places here, and the abstract ideas become graphic on notebook pages, in newspaper headlines, or through the art they have engendered. The apparatus of physics, usually eminently illustratable, has become ever more impressive (and expensive!), as Henri Becquerel's simple uranium ore and photographic plate and C. T. R. Wilson's early cloud chamber gave way to huge experimental fusion machines and particle detectors.

Chemistry, like astronomy, has been heavily invaded by physicists in the twentieth century. Nearly half of the American Chemical Society's posterlike time line for the century is devoted to the understanding of the nuclear atom or the discovery of heavy elements in physical laboratories, and it is easy to track down images of the equipment that made these advances possible. But pictures from the scientific world of molecules are far more difficult to locate. For industrial chemistry, photographs come readily to hand showing factories or useful new products: nylon, Plexiglas, synthetic rubber, Teflon—materials so ubiquitous that we almost take them for granted. But try to find an honest, unposed picture of a chemist at work in the laboratory on Nobel-quality research, and you will begin to appreciate the challenge faced in preparing this *Album*.

Geologists have two colossal achievements to their credit since 1900: the absolute dating of rocks and strata (with some help from the physicists) and the observational evidence for the theory of continental drift and plate tectonics (with some help from the deep-sea oceanographers, who may have been geologists in disguise). Some of the ancillary phenomena of the movement of plates—earthquakes and volcanoes—are visually spectacular, so they are included here. Graphs of the seafloor must be an acquired taste, but they are a visible part of the evidence and have been pasted into the scrapbook.

Marching hand in hand with science is technology. Science has made some of that technology possible, and even more so has technology made much of the science possible. "Talkie" movies demanded improved photocells that made photoelectric photometry a practical reality for astronomers, just as sniperscopes developed during the Vietnam war put infrared astronomy on the map. Is an atom smasher or a space-borne laboratory technology or science? What about airplanes and rockets, television and computers? This scrapbook's pictorial bias slants against technology, but the borderline continually blurs. The Wright brothers' flight was too remarkable to omit. Radio

brought in electronics, and high-speed computers made possible interplanetary exploration, the discovery of the W and Z nuclear particles, the mapping of enzyme topology, and the calculation of stellar evolution, to name but a few examples. Consequently, multiple images of radio technology and computers will be found on these pages. Indeed, the "electronic revolution" of data acquisition and data handling largely characterizes the science of the century's final decades.

Some will find mathematics strangely absent here. A handmaiden of science, mathematics is not science; it is an interlocked logical structure lacking the experimental or observational interface with the natural world that characterizes science, and thus is intentionally ruled out of bounds for our volume.

Science in the twentieth century has involved far more workers and far more financial support than all previous centuries combined. Hence, for sheer lack of space, the twentieth-century volumes of the *Album of Science* series necessarily skip and choose far more than the previous ones. It is no wonder that some topics and a few favorite pictures may seem mysteriously omitted.

The mention of workers and financial support reminds us that there is more to the physical sciences than simply the study of physics, chemistry, astronomy, geology—there are also the organizational structures, the prizes and grants, the journals and referees, international competition and cooperation, the popularization of science and its public image. Hence the images here of the explosive growth of the *Physical Review*, the Solvay conferences, the science pavilions of the international expositions and science museums around the world. As always, the selection is not comprehensive, but it gives the flavor of the intricate relationship between science and society.

You will find here a stimulating and evocative set of images that convey the sweep and excitement of nearly a century of work in the physical sciences. By its nature such a set can never be "complete" and must inevitably represent a somewhat personal and even idiosyncratic vision, a vision at times expanded by serendipitous pictorial discoveries but at other times fenced in by the particular illustrations that could actually be acquired. In the process of preparing this *Album* my own vistas have broadened, and I have been educated, occasionally baffled, and sometimes amused. I hope those who open this book will enjoy some of the same experiences.

Part One

THE AGE OF RUTHERFORD, EINSTEIN, AND BOHR

Seizième année. — N° 779.   Huit pages : CINQ centimes   Dimanche 10 Janvier 1904.

Le Petit Parisien

SUPPLÉMENT LITTÉRAIRE ILLUSTRÉ

TOUS LES JOURS
Le Petit Parisien
(Six pages)
5 centimes

CHAQUE SEMAINE
LE SUPPLÉMENT LITTÉRAIRE
5 centimes

DIRECTION: 18, rue d'Enghien (10ᵉ), PARIS

ABONNEMENTS
—
PARIS ET DÉPARTEMENTS:
12 mois, 4 fr. 50. 6 mois, 2 fr. 25

UNION POSTALE:
12 mois, 5 fr. 50. 6 mois, 3 fr.

UNE NOUVELLE DÉCOUVERTE. — LE RADIUM
M. ET Mᵐᵉ CURIE DANS LEUR LABORATOIRE

2. The Curies in their laboratory. Following Henri Becquerel's discovery in 1896 that uranium emitted radiation, Marie Curie began looking for similar substances. Her husband, Pierre, joined in the research, and in 1898 they announced the discovery of polonium and radium in pitchblende. The discovery of radioactivity earned the Curies, along with Becquerel, the Nobel Prize in physics in 1903. Despite their years of achievement, the Curies still had only a meager laboratory.

1

*R*adioactivity
and the *N*ucleus

Twentieth-century science began in 1896. In February of that year the French physicist Henri Becquerel discovered that a compound of uranium (potassium uranyl sulfate) emitted radiation that could penetrate paper opaque to light and could darken a photographic plate. This phenomenon was later called radioactivity by Marie and Pierre Curie, who devoted many years to studying it. The discovery of radioactivity—a phenomenon that generally involves the transmutation of one chemical element into another—was the first of many shocks to the comforting nineteenth-century belief in the stability of the physical world, though it followed by nearly four decades Charles Darwin's revolutionary hypothesis that one biological species could be transmuted into another. Soon mass, energy, time, space, and determinism would also lose their privileged positions in science, and even the objective existence of a reality apart from human observation would be cast into uncertainty.

Becquerel chanced upon radioactivity while seeking substances that give off X rays, which had been discovered a few months earlier by the German physicist Wilhelm Röntgen. X rays, emitted by certain substances when exposed to a beam of cathode rays, created a great sensation among the public because they passed through human flesh and could produce a "picture" of the skeleton together with other incidental solid objects—bullets, swallowed pins, etc. But the idea that one kind of ray could be transformed into another was not new to the physics of the time. X rays, myste-

rious as they seemed at first, would not by themselves have produced a scientific revolution. Radioactivity was different. Here was a substance emitting radiation spontaneously, without any external stimulus, and without undergoing any change itself—or at least that was what people thought. In fact, radioactivity was heralded in the popular press as an amazing new source of energy.

After Becquerel made the initial discovery, the pioneering role in radioactivity research passed to Marie Skłowdowska Curie, a young Polish woman who had come to Paris to study science and had married Pierre Curie, an older French physicist. For her doctoral dissertation she decided to search for a highly radioactive substance that seemed to be present in very small amounts in some uranium-bearing minerals, especially pitchblende. Pierre put aside his own research to help his wife in the enormous task of separating and studying two new elements—polonium and radium. By 1902, Marie had completed the first stage of the work by establishing the atomic weight of radium and determining its other properties. The next year she received her doctorate in physics and shared the Nobel Prize with Pierre and Becquerel.

Marie Curie's work with radium provided the scientific community with a source of intense radioactivity. Other scientists, with better facilities at their disposal than the Curies had (even after the Nobel Prize), quickly made a number of major discoveries and proposed bold new theories about the nature of radioactivity. Ernest Rutherford, a native of New Zealand who worked at McGill University in Montreal and later at Manchester and Cambridge universities in England, emerged as the leader in this fast-moving field.

Rutherford first showed that uranium emitted two different kinds of radiation, which he called alpha rays and beta rays. Alpha rays were much less penetrating—a piece of paper was enough to stop them. Beta rays went through the paper but were stopped by a thin sheet of aluminum foil. A third kind of radiation, discovered by the French physicist Paul Villard, was even more penetrating; Rutherford called it gamma rays. In 1903, Rutherford concluded that alpha rays, which had been shown to have positive electric charge, were identical to doubly ionized helium atoms, while beta rays, known to have negative charge, were the same as the "electron" discovered in 1897 by J. J. Thomson. Gamma rays, with no charge, were (like X rays) electromagnetic radiation of very high frequency.

Rutherford and his colleague Frederick Soddy proposed that after a uranium atom emitted an alpha ray it was no longer a uranium atom but had broken up into helium (alpha ray) and thorium. Further research indicated that thorium is also radioactive; there is a long chain of successive radiative transmutations, ending up with lead, which is stable.

The Rutherford-Soddy hypothesis required abandoning the doctrine that matter comes in a large number of qualitatively different kinds, or elements. It now seemed possible that there existed only three or four distinct particles or kinds of matter, from which everything was constructed.

Rutherford took a major step toward understanding the structure of the atom in 1911. On the basis of an experiment by his assistant Hans Geiger and a student, Ernest Marsden, he proposed that the atom consisted of a small dense nucleus with a positive electric charge; most of the rest of the atom was empty except for negatively charged electrons that moved around the nucleus.

In 1913 a Dutch scientist, Antonius van den Broek, suggested that the electric charge on the atomic nucleus was exactly equal to the "atomic number" in Mendeleev's periodic table of the elements. This suggestion was shown to be correct by Rutherford's colleague Henry Moseley on the basis of a systematic survey of X-ray spectra. The atomic number would also be equal to the number of electrons needed to balance the nuclear charge and make a neutral atom. Thus it became possible to explain the chemical properties of elements in terms of a simple physical property: the charge on the nucleus. The Dane Niels Bohr, who in 1913 proposed a simple model to explain how the electrons moved around Rutherford's nucleus, was the first to show how this might be done.

It was also possible to make sense out of radioactive transmutations, by assuming that the total nuclear charge is conserved. Thus, when a uranium atom, with nuclear charge 92, decays to thorium, nuclear charge 90, it emits an alpha ray, which is a helium nucleus with charge 2. When this thorium atom decays to protactinium, with nuclear charge 91, it emits a beta ray with charge -1 [since $90 = 91 + (-1)$]. Using the charge conservation rule, Rutherford and his colleagues were able to identify even the short-lived radioactive substances and assign them to places in the periodic table. But this soon produced a puzzle: in several cases two or more substances had to go into the same place even though their physical properties seemed to be different. In Chapter 2 we shall see how this problem was solved.

Stephen G. Brush

4

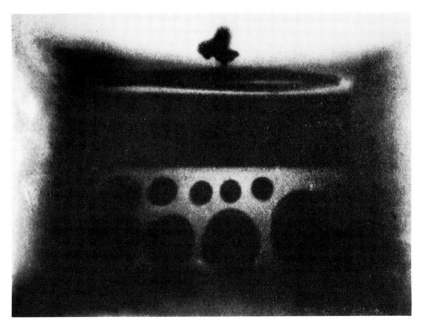

3. Röntgen's laboratory. The discovery of X rays in 1895 by the German physicist Wilhelm Röntgen presaged the birth of modern physics and earned Röntgen, in 1901, the first Nobel Prize in physics. This 1923 photograph shows the laboratory at the Würzburg Physical Institute where Röntgen discovered the penetrating rays.

4. First X-ray image. In studying the previously unknown rays emitted from the glass wall of a cathode-ray tube, Röntgen discovered that with them he could "photograph" the metal weights of a balance inside their closed oak case. Soon he obtained images of other concealed objects, such as living bones.

5

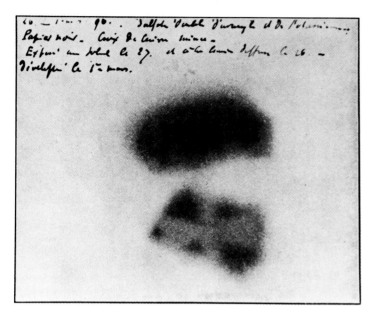

5. Discovery of radioactivity. Röntgen's discovery of X rays was quickly pursued. Henri Becquerel speculated that all forms of luminescence might be accompanied by X rays. One day early in 1896 he wrapped up a photographic plate, put a phosphorescent uranium salt on it, and exposed the whole thing to the sun. The developed plate showed a black outline. It seemed that the substance emitted X rays when it fluoresced in the sunlight. He tried to repeat the experiment, but there was not enough sunshine, and so he put his wrapped plates and uranium compound in a dark drawer for a few days. When he developed the plates, he found, to his surprise, an intense exposure. Becquerel had thereby discovered radioactivity. Shown here is one of the plates from the drawer, with Becquerel's notation.

6. Father of atomic physics. With his discovery of radioactivity in 1896, Becquerel began a line of investigation that was to play a major role in twentieth-century science, although at the time it did not seem as impressive as Röntgen's X rays. Becquerel was partly working within a family tradition, for his father and grandfather were also physicists, and the Frenchman said they deserved some credit for his discovery. Here we see him in the laboratory with a large electromagnet, illustrating the interests of his forebears as well as much of his own research.

7. Geiger and Rutherford. Hans Geiger, who developed techniques for detecting and counting charged particles, worked for several years under the supervision of Ernest Rutherford (*right*) at the University of Manchester in England. Here we see them in 1912 counting alpha particles from radioactive atoms. Four years earlier, using a proto-"Geiger counter," they proved that alpha particles are doubly charged.

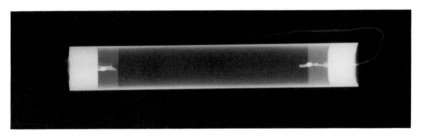

8. X-ray view of a counting tube. In the late 1920s Geiger and Walther Müller developed the Manchester counter into a practical device. The "copper counting tube" shown here was made in 1928. A particle entering the tube ionized the gas inside, producing an electrical pulse.

9. Early Geiger-Müller counters. On the laboratory bench are the counters used around 1934 in Rome, which had developed into a major center of physics research. There Enrico Fermi, Emilio Segrè, and others were investigating "artificial" radioactivity induced by bombarding atoms with neutrons.

RADIO-ACTIVITY:

AN ELEMENTARY TREATISE,

From the Standpoint of the Disintegration Theory.

BY

FREDK. SODDY, M.A.,

LECTURER ON PHYSICAL CHEMISTRY AND RADIO-ACTIVITY
IN THE UNIVERSITY OF GLASGOW.

WITH FORTY ILLUSTRATIONS.

COPYRIGHT.

ENGLAND:
"THE ELECTRICIAN" PRINTING & PUBLISHING COMPANY, LTD.,
SALISBURY COURT, FLEET STREET, LONDON.

1904.

from air, allowed to flow into B. The seal at the dotted line N was then made. C is a tube in which is a thin copper spiral wire which can be heated by an electric current. D is a phosphorous pentoxide tube to absorb moisture. E is a capillary U tube which is cooled in liquid air during the experiment. The emanation and any CO_2 present are condensed here and prevented from entering the spectrum tube.

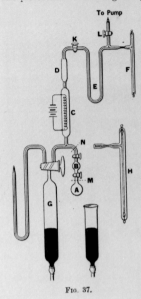

FIG. 37.

The tap L is connected to the mercury pump not shown. F is the spectrum tube shown half full-size at H. The copper spiral is first partially oxidised by filling the tube with oxygen from the burette G and keeping the spiral at dull redness by a current. The whole apparatus is thoroughly exhausted and all taps closed. Water is now admitted from B into A, the

10. Helium from radium. The British chemist Frederick Soddy worked with Rutherford on the disintegration theory of radioactivity and then demonstrated with William Ramsay, using the apparatus shown here, that helium was a disintegration product of radium. In 1910 he advanced the concept of isotope, for which he won the 1921 chemistry Nobel Prize.

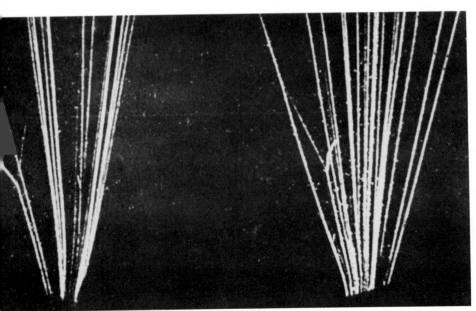

11. Artificial atomic transmutation. For this stereoscopic photograph a source of alpha particles was placed in a Wilson cloud chamber filled with nitrogen. One of the alpha particle tracks branches in two directions, indicating that a collision has occurred; Rutherford realized that a proton had been ejected. Analysis by P. M. S. Blackett showed that a nitrogen nucleus had been converted into oxygen. The ancient alchemical dream of transmuting elements was thus finally fulfilled.

12. Wilson cloud chamber.
Developed by the Scottish physicist
C. T. R. Wilson at Cambridge University's Cavendish Laboratory between 1896 and 1912, the cloud
chamber provided a spectacular way
to observe the paths of charged
particles. A fast-moving alpha particle or proton will produce ions
along its path, around which tiny
water droplets can condense, making the track visible. The track can
then be observed or photographed
through the chamber's glass top.

13. Rutherford's laboratory. In 1919, Rutherford moved from Manchester
to become J. J. Thomson's successor as the director of the Cavendish Laboratory. Shown below is his research room in the early 1920s.

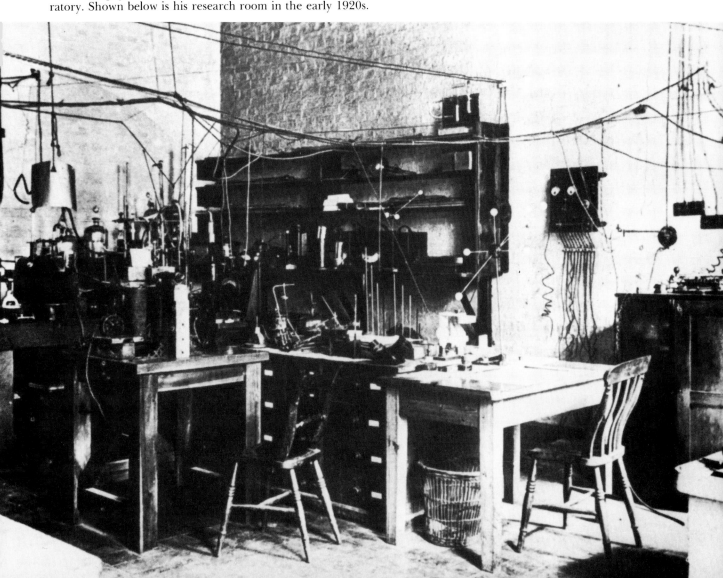

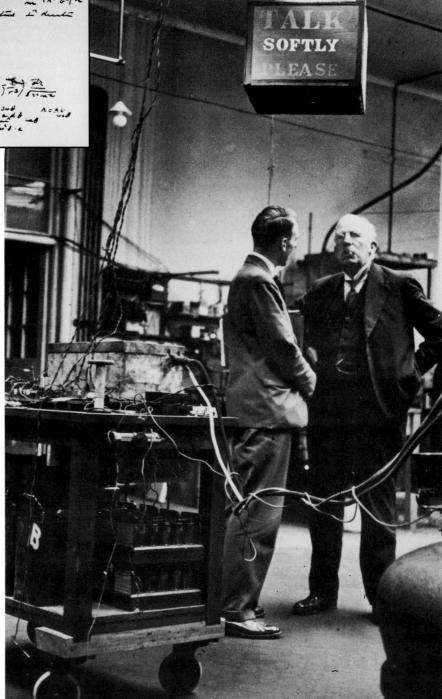

14. Toward the nuclear atom. This page from Rutherford's notes, written in Manchester in 1911, represents his first serious thoughts on the theory of an atomic nucleus. The right-hand sketch depicts J. J. Thomson's "plum pudding" atom with negatively charged electrons scattered through a positive fluid. The left-hand sketch and calculation show the deflection a charged particle might undergo in traversing the vicinity of a tiny charged nucleus.

15. The boss of the Cavendish. Rutherford, spreading tobacco ash and noise, talks with Cavendish researcher J. A. Ratcliffe. The particle-counting device on the lab bench behind Ratcliffe was extremely delicate—hence the homemade sign above. Rutherford had a booming voice and was constitutionally incapable of talking softly. C. E. Wynn-Williams, who developed the device, turned on the illuminated sign before taking the picture.

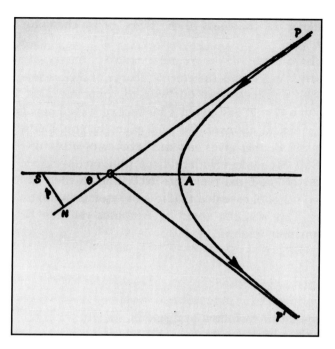

16. Deflected alpha particle. Experiments at Manchester showed that alpha particles could be deflected at unexpectedly large angles as they shot through a foil of gold atoms. In a 1911 paper Rutherford analyzed this "alpha-particle scattering." He calculated that a positively charged alpha particle passing close to a positively charged nucleus would be repelled along a hyperbolic path such as the one shown in the diagram. These findings lent support to the hypothesis of an atomic nucleus.

17. The Rutherfords and the Bohrs. The Danish physicist Niels Bohr worked briefly under Rutherford at Manchester in 1912. There he laid the foundations for his greatest achievement in physics, the theory of atomic structure. It was Bohr who first grasped the far-reaching implications of Rutherford's nuclear model of the atom. Thus began a lifelong friendship. The picture was taken around 1930 by the Australian physicist Marcus Oliphant, whose wife sits between Mrs. Rutherford and Mrs. Bohr.

18. Richards in his laboratory. In 1914, Theodore William Richards became the first American chemist to receive a Nobel Prize. The award recognized his determination of the atomic weights of twenty-five elements. Richards found serious errors in some of the accepted figures, and he also showed that lead from radioactive materials has a different atomic weight from ordinary lead. More than sixty noted twentieth-century chemists were products of Richards's teaching at Harvard.

2

*A*tomic *C*hemistry

The electron was the first subatomic particle to be found. Its discovery was not a single historical event but rather a combination of experiments, debates, and theoretical speculations in the last part of the nineteenth century, culminating in the work of the British physicist J. J. Thomson in 1897. Thomson established the view that the so-called cathode rays seen in electrical discharges in vacuum tubes were streams of discrete particles with negative electric charge and small but finite mass. The opposing theory was that cathode rays were some form of electromagnetic waves or disturbance in the light-transmitting "ether" that was thought to fill all space. For a while it was accepted that Thomson had proved that electrons were definitely particles rather than waves, but in the 1920s physicists began abandoning the assumption that "particle" and "wave" are mutually exclusive designations. Curiously, although Thomson in 1906 had been awarded the Nobel Prize in physics for proving the particle nature of the electron, in 1937 his son George Thomson shared the Nobel Prize for showing that the electron has a wave nature.

J. J. Thomson was able to measure the charge-to-mass ratio of the electron. Soon afterward, others in his laboratory at Cambridge University attempted to measure its charge directly. The definitive research, however, was undertaken by an American physicist, Robert A. Millikan, in 1906. Millikan found a way to balance the electric and gravitational forces on a single oil drop. His exper-

iments, completed in 1914, showed that the electric charge on the oil drops always came in integer multiples of a certain basic charge, known as the electronic charge. Combining this value for the charge with Thomson's measurement of the charge-mass ratio led to an estimate of the electron's mass, about 9.1×10^{-31} kilogram.

The electron's mass is thousands of times smaller than the mass of an atom. Thus, if electrons are constituents of atoms, either there must be thousands of electrons in an atom or else atoms must also contain more massive types of particles. The first such particle to be discovered was the hydrogen nucleus, identified by W. Wien and J. J. Thomson in cathode rays and named the proton by Rutherford. It has a positive electric charge with the same magnitude as that of the electron, and a mass of about 1.67×10^{-27} kilogram; in other words, it is about 1,800 times as massive as the electron.

By that time scientists had a fairly definite conception of the atomic structure of matter. At the end of the nineteenth century a few skeptics raised philosophical doubts about atoms on the ground that science should not deal with entities that could not be directly observed. J. J. Thomson and his followers dispelled the doubts by showing that one could make quantitative measurements of atomic particles such as electrons even though it was not yet possible to "see" them. Their trails could be seen. The celebrated cloud chamber constructed by the Scottish physicist C. T. R. Wilson made visible tracks of water droplets condensed along the path of a charged particle.

Another argument for the existence of atoms came from Albert Einstein's theory of Brownian movement, published in 1905 and quickly confirmed by the French physicist Jean Perrin. The theory depended on statistical fluctuations in the motions of atoms striking a microscopic particle suspended in a fluid and could be quantitatively tested by observations of the particle. Perrin's experiments led to values for the size and mass of an atom that were consistent with those obtained by other methods. In a sense it was Perrin who finally established the reality of atoms to the satisfaction of the scientific community. (The first person to actually "see" an atom was Erwin W. Müller, in 1955, with the field-ion microscope he perfected.)

In the years after 1910 these discoveries about electrons and atoms were quickly combined with the findings on radioactivity and the nucleus to construct a general theory of atomic structure. Two fundamental properties were known: the atomic number, shown by Henry Moseley to be proportional to the electric charge on the nucleus, and the atomic weight, accurately determined for several more elements by the American chemist T. W. Richards. When all the known radioactive substances were assigned to places in the periodic table, it turned out that several substances with different atomic weights had to be put in the same place, with the same atomic number. Frederick Soddy introduced the term "isotope," meaning "same place," for atoms with the same atomic number but different atomic weights.

Since isotopes of an element could not be distinguished by chemical methods, it was difficult to determine their atomic weights. The mass spectrograph, invented by Francis W. Aston at Cambridge University in 1919, allowed direct determination of the atomic weights of isotopes by using a combination of electric and magnetic fields.

An important clue to the structure of the atomic nucleus was uncovered in 1920 when Aston announced that all atomic weights are close to integers if the figure for oxygen is taken as 16. (The reason why they are not exactly integers will be explained in connection with Einstein's theory of relativity in Chapter 3.) The British chemist William Prout had suggested a century earlier that the atomic weights should all be exact multiples of hydrogen, but accurate determinations showed significant deviations from this rule. The most serious discrepancy was provided by chlorine, whose atomic weight was 35.46. Aston found that there are two isotopes of chlorine, with atomic weights approximately 35 and 37. Since isotope 35 makes up about 75 percent of the chlorine atoms in nature while isotope 37 accounts for the remaining 25 percent, the average atomic weight is about 35.5. In most cases the isotopes of an element are not very different in mass. The major exception is hydrogen, whose isotope deuterium, discovered by the American chemist Harold Urey in 1931, is twice as massive as the most abundant isotope.

If Prout's hypothesis were valid, one might expect that all atomic nuclei would be compounds of the nucleus of hydrogen—the proton. The atomic weight would be the number of protons, while the atomic number would be the net positive charge remaining after some of the protons had been neutralized by electrons. This was indeed the theory of nuclear structure generally adopted in the 1920s, but as we shall see in Chapter 12, it had to be changed.

The year 1911, in which Rutherford proposed his nuclear model of the atom, also saw two major discoveries in low-temperature physics by the Dutch physicist Heike Kamerlingh Onnes. In 1908, Kamerlingh Onnes had succeeded in liquefying helium by cooling it to -268.9° Celsius, just 4.2 degrees above absolute zero on the Kelvin temperature scale. In 1911 he noticed that liquid helium reached a maximum density at about 2.2 degrees above absolute zero; it expanded when cooled below that temperature. Scientists later found that at this temperature helium undergoes a transition, called the lambda transition (because the graph of heat capacity looks like the Greek letter lambda), to a peculiar physical state. In this state helium is "superfluid"—it can easily pass through microscopic cracks or capillary tubes.

Kamerlingh Onnes's other 1911 discovery was superconductivity: he found that when he cooled a metal to a very low temperature, its electrical resistance suddenly dropped to zero. The study of other strange properties of superconductors, such as magnetic levitation, eventually became a major area of research in twentieth-century physics, especially after scientists learned how to concoct substances that are superconducting at much higher temperatures.

Stephen G. Brush

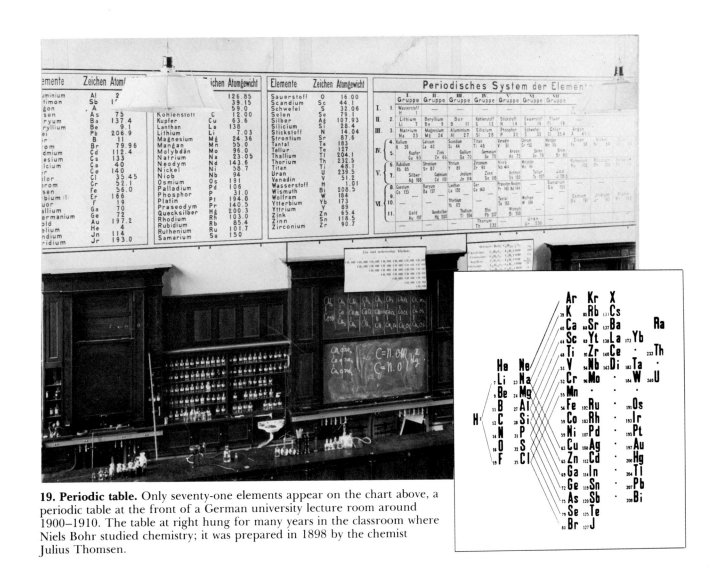

19. Periodic table. Only seventy-one elements appear on the chart above, a periodic table at the front of a German university lecture room around 1900–1910. The table at right hung for many years in the classroom where Niels Bohr studied chemistry; it was prepared in 1898 by the chemist Julius Thomsen.

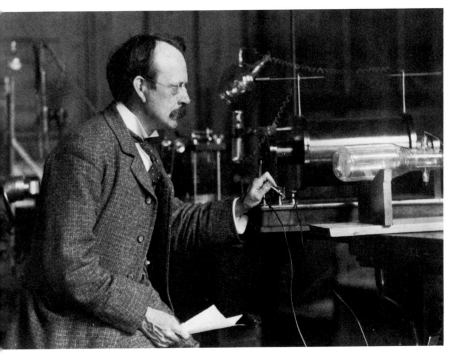

20. Thomson and the electron. This late nineteenth-century photograph of J. J. Thomson giving a lecture demonstration of the cathode-ray tube was taken at about the time he discovered the electron. Thomson settled an old controversy over the nature of cathode rays by confirming that they are particles. He found that the rays were the same regardless of the cathode material or the gas in the tube.

21. The oil drop experiment. With this apparatus the American physicist Robert Millikan refined the procedure for measuring the charge of the electron. Using charged droplets of oil and an electric field to balance the force of gravity, he showed, just before World War I, that the charge on a drop was always an integral multiple of a basic value—evidence that electrons were fundamental particles having a specific charge.

22. Brownian motion and the existence of atoms. Fine particles suspended in a liquid are in continual random motion. The two pictures—a photomicrograph (*left*) showing particles of gum resin in a drop of water and a drawing (*right*) of the path of a single particle—are from the French physicist Jean Baptiste Perrin. His studies confirmed Albert Einstein's theory of Brownian motion and provided irrefutable proof that atomic theory's atoms and molecules were not just convenient fictions.

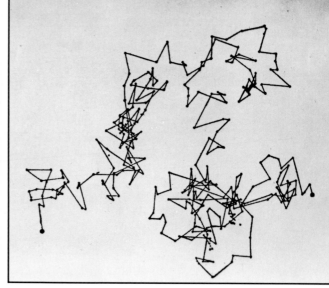

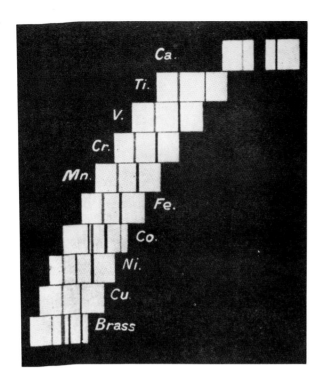

23. X-ray spectra and atomic number. In a brilliant series of measurements of X-ray spectra, the British physicist H. G. J. Moseley showed that key properties of chemical elements are determined by the atomic number (the charge on the nucleus), rather than by the atomic weight. This discovery, called by Millikan one of the most illuminating results in the history of science, made it possible to test the periodic table for completeness. Moseley published these spectra of the so-called K series in 1913: arranged in order of X-ray frequency, they also place the elements in order of atomic number.

24. Moseley at Oxford. The young physicist is shown here in the Balliol-Trinity laboratory around 1910, the year he took up a position with Rutherford at the University of Manchester. A scientist of great promise and marked achievement, Moseley died in 1915 at the age of twenty-seven, killed in action during the British expedition to the Dardanelles.

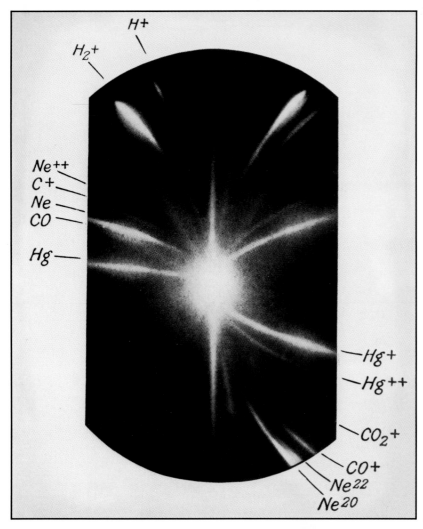

25. Telltale parabolas. J. J. Thomson worked extensively with "positive rays"—positively charged particles that travel opposite to the negative cathode rays. When deflected by electric and magnetic fields, they produced parabolic tracks representing a constant ratio of charge to mass. Did this mean that all atoms of a given substance had the same mass? Thomson in 1912 found two parabolas for neon, representing mass 20 and 22. Not until 1920 did Thomson's sometime assistant Francis Aston prove by separation that these really were two different isotopes of neon.

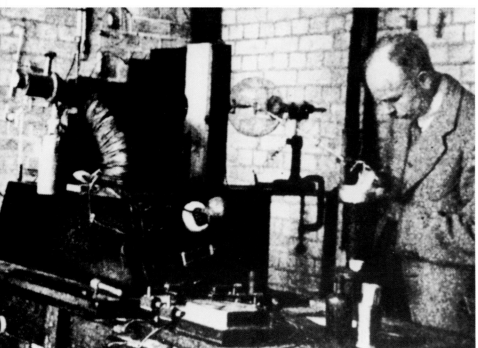

26. Aston and his first mass spectrograph. The notion that the two parabolas obtained for neon represented two different isotopes of the same element was confirmed by Aston at the Cavendish Laboratory. In 1919 he built a new kind of apparatus, which he called a mass spectrograph and which provided not only much better resolution but greater accuracy than Thomson's parabolas. Aston subsequently developed larger and more precise mass spectrographs; by the time of his death in 1945 he had discovered three-fourths of the naturally occurring isotopes.

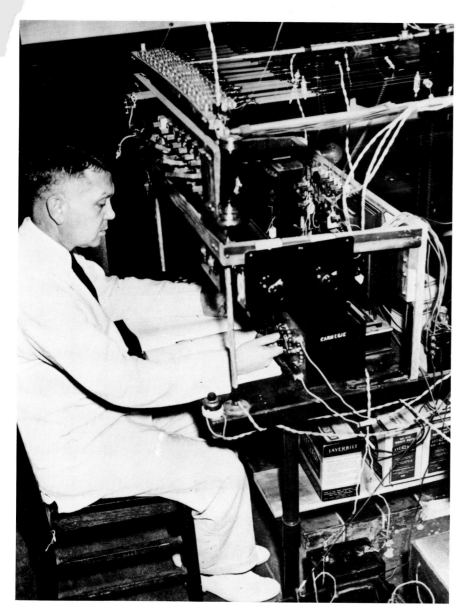

27. Urey at the controls. With this mass spectrometer the American chemist Harold C. Urey and colleagues in 1931 discovered the stable isotope of heavy hydrogen known as deuterium. Urey's group was the first to achieve a nearly complete separation of one isotope from another. With a nucleus of one proton and one neutron, deuterium is about twice as heavy as the most common form of hydrogen. Deuterium was subsequently used in hydrogen bombs, and the compound deuterium oxide, or heavy water, found application as a moderator in nuclear reactors.

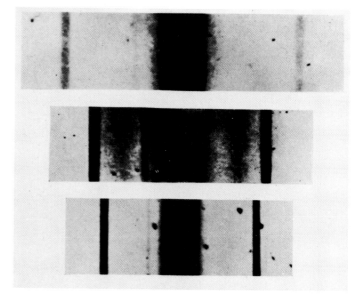

28. Telling spectra: discovery of deuterium. These enlargements of the three brightest spectral lines of hydrogen were published in Urey's discovery paper in 1932. At the center of each spectrum is a heavily overexposed line of the most common isotope of hydrogen. To the left (that is, on the violet, or high-frequency, side) of each of these intense lines is the faint line due to deuterium. The symmetrically placed outliers are ghost lines that arise from errors in the spectrograph's grating.

29. Dewar. The British chemist and physicist James Dewar was one of the pioneers of low-temperature physics. Beginning his work in cryogenics in 1877, he succeeded by 1885 in producing liquid air and liquid oxygen in quantity. In 1892 he invented a vacuum-jacketed flask, the most important device for storing low-temperature materials; this double-walled flask (with vacuum between the walls) became known as the Dewar, or Thermos, flask. He obtained liquid hydrogen in 1898 and solidified it the following year. Dewar cooled the solid hydrogen to just 13 degrees above absolute zero, the lowest temperature achieved to that time.

30. Kamerlingh Onnes and van der Waals. These two Dutch scientists won the Nobel Prize in physics in 1913 and 1910, respectively. Heike Kamerlingh Onnes (*left*) devised a superior apparatus for achieving low temperatures and in 1908 succeeded in liquefying helium, edging out Dewar in the race toward absolute zero. The photograph, showing the helium liquefier, was taken in Kamerlingh Onnes's laboratory at the University of Leiden in 1911, the year he discovered superconductivity—the lack of electrical resistance. He observed the phenomenon in mercury at temperatures near absolute zero. Johannes van der Waals made major contributions to the theoretical description of liquids and gases. Both Kamerlingh Onnes and Dewar built on van der Waals's research in their liquefaction of gases.

31. Kamerlingh Onnes's liquefier. Kamerlingh Onnes liquefied helium with this apparatus in 1908. The schematic is his. Beginning at tube C_a, the helium passed through various refrigerating tubes and coils and finally expanded through cock M_1. This expansion further cooled the helium. If sufficiently cool, the helium collected in liquid form in the lower part of the vacuum glass E_a.

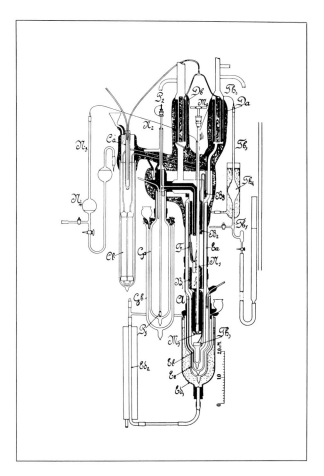

32. Kapitsa's liquefier. The Soviet physicist Pyotr (Peter) Kapitsa worked with Rutherford in Cambridge from 1921 to 1934 but on a visit to the U.S.S.R. was compelled to remain there. Kapitsa became director of the Institute of Physical Problems, where this photograph of the Cryogenic Hall was taken in 1939. Working in low-temperature physics, he devised an engine to produce abundant liquid helium and thereby discovered superfluidity, an unexpected property of the coldest form of liquid helium.

33. Einstein. Here we see Albert Einstein in the Swiss Patent Office in Bern, where he worked from 1902 to 1909. The photograph was taken about 1905, the year in which he published not only his special theory of relativity and the equivalence of mass and energy but also a theoretical explanation of Brownian motion and an analysis of the photoelectric effect in terms of quanta of light that exhibit particlelike behavior (for which he won the 1921 Nobel Prize in physics).

3

*R*elativity

One of the great controversies of the seventeenth century concerned the relation between the earth and the celestial bodies. Do the sun, planets, and stars move around a fixed earth, or does the earth move with the planets around a fixed sun while the stars are at rest? The debate assumed, of course, that it makes sense to say that something is "really moving" or "really at rest."

Isaac Newton, in his *Principia* (1687), asserted the reality of absolute space, time, and motion. In other words, there is a unique coordinate system with respect to which every object in the universe is either moving or at rest, and there is a unique time scale so that, for any two events *A* and *B*, either (1) *A* and *B* are simultaneous or (2) *A* comes before *B* or *B* comes before *A*. These propositions came to seem so obvious by the nineteenth century that scientists scarcely felt it necessary to state them explicitly. Newton had proved that absolute rotational motion could be detected through the effects of centrifugal force, and he pointed out that the earth's equatorial bulge (its deviation from perfectly spherical shape) is a consequence of its rotation. Later, phenomena such as the patterns of winds and ocean currents and the motion of the Foucault pendulum were also ascribed to the earth's rotation. When stellar parallax—the apparent shift in position of certain stars as seen from the earth at different times of the year—was discovered in the nineteenth century, it was thought to be a result of the earth's motion around the sun.

All such observations showed only that absolute rotation or acceleration could be detected. They gave the impression that absolute space, time, and linear motion exist, but this impression was misleading.

When the wave theory of light was adopted by scientists in the 1820s, it appeared to require the existence of an ether filling all space, in order that vibrations could be transmitted from one place to another. Presumably light would travel at constant speed relative to the ether; hence if the earth moves through the ether, the speed of light measured in the laboratory should be different in directions parallel or perpendicular to the earth's motion. Following a suggestion by the British physicist James Clerk Maxwell, A. A. Michelson and E. W. Morley in the United States tried to measure the earth's motion relative to the ether by observing the interference of two parts of a beam of light sent along different paths.

The Michelson-Morley experiment, completed in 1887, showed no measurable difference in the speed of light in different directions, although the apparatus was capable of detecting the difference that should have been produced by the earth's motion around the sun. Thus, it seemed impossible to detect the earth's absolute motion by this method. This result could not be plausibly explained by existing physical theory; G. F. FitzGerald in Ireland and H. A. Lorentz in the Netherlands did propose a rather artificial hypothesis: the solid arm of the interferometer that determined the length of the light path contracted in the direction of the earth's motion by an amount just sufficient to cancel out the expected effect.

Although the negative result of the Michelson-Morley experiment was later to provide one of the strongest justifications for his theory of relativity, Albert Einstein did not invent the theory for the purpose of explaining the experiment. Instead, he was primarily concerned with the fact that Maxwell's electromagnetic theory treated the interaction of electric and magnetic fields inconsistently: the equations were different depending on which coordinate system was used, even though the results came out the same. Einstein found that a more consistent formulation of the theory could be developed by abandoning the assumption that absolute space, time, and motion exist. In a famous paper published in 1905, he replaced that assumption by two postulates: (1) the laws of nature are the same for observers in any frame of reference, and (2) the speed of light is the same for all such observers. ("Frame of reference" here means an inertial frame, that is, one in which the law of inertia is valid; a rotating or accelerating coordinate system would not qualify.)

From his two postulates Einstein deduced that an observer in one frame of reference would find from his own measurements that objects in another frame are contracted in accordance with the formula proposed earlier by FitzGerald and Lorentz; the effect is reciprocal. Also, an observer in each frame would find that time intervals between events in the other frame are expanded, or dilated, by a similar factor—that is, clocks in the other frame would appear to be running more slowly. Finally, events that are simultaneous for an observer in one frame may not be so for an observer in another frame.

In another paper published in 1905, Einstein showed that the observable mass of any object should increase as it goes faster, with the mass becoming infinite at the speed of light. Thus, the speed of light is the maximum possible speed.

Closely connected with the mass-increase effect is Einstein's famous formula $E = mc^2$: mass (m) and energy (E) are no longer separately conserved but can be interconverted. In particular, the total mass of a nucleus is not precisely equal to the sum of the masses of its constituents, because of the energy corresponding to intranuclear forces. This is why the masses of isotopes are not exactly integers. Nuclear reactions in which the total mass of the products is less than that of the reagents will release energy. One example is the fission of heavy nuclei such as uranium or plutonium in the atomic bomb or in a nuclear reactor; another is the fusion of light nuclei such as hydrogen in the hydrogen bomb or in the interiors of stars.

These assertions are consequences of the "special" theory of relativity, in which gravitational forces are not involved and the reference frames of the observers are not accelerated. In 1907, Einstein took the first step toward a more general theory by proposing his principle of equivalence: at any given place a gravitational field is equivalent to an acceleration. This suggests that gravity may simply be a property of space and time. Einstein later derived a set of equations in which gravity is related to the "curvature" of a consolidated space-time coordinate system. Here he made use of the theory of non-Euclidean geometry that was developed by the German mathematician G. F. B. Riemann in the nineteenth century. In Riemann's system the universe is finite but has no boundary; if you keep on traveling in the same direction, you eventually come back to where you started, just as if you were on the surface of a sphere.

From this "general" theory of relativity Einstein predicted that light from a star would be bent by the sun's gravitational field. The prediction's confirmation in 1919 quickly made him famous. Einstein also succeeded in solving a long-standing problem in planetary astronomy: the gradual shift in the orbit of the planet Mercury. He showed that this shift was due to the fact that the mass of the sun as observed on a distant planet is different from the mass observed on a nearby planet, because of the binding energy of the gravitational force—just as the mass of a particle inside a nucleus is different from its mass determined when it is outside a nucleus. So a planet like Mercury, whose distance from the sun varies significantly as it traverses its orbit, will experience a variable solar mass.

The spectacular success of relativity theory apparently persuaded Einstein that the true path to unlocking the secrets of nature lay in the realm of abstract mathematical thought. But his efforts to develop an even more general theory that would encompass all forces and particles never produced a satisfactory result.

Stephen G. Brush

3. Zur Elektrodynamik bewegter Körper; von A. Einstein.

Daß die Elektrodynamik Maxwells — wie dieselbe gegenwärtig aufgefaßt zu werden pflegt — in ihrer Anwendung auf bewegte Körper zu Asymmetrien führt, welche den Phänomenen nicht anzuhaften scheinen, ist bekannt. Man denke z. B. an die elektrodynamische Wechselwirkung zwischen einem Magneten und einem Leiter. Das beobachtbare Phänomen hängt hier nur ab von der Relativbewegung von Leiter und Magnet, während nach der üblichen Auffassung die beiden Fälle, daß der eine oder der andere dieser Körper der bewegte sei, streng voneinander zu trennen sind. Bewegt sich nämlich der Magnet und ruht der Leiter, so entsteht in der Umgebung des Magneten ein elektrisches Feld von gewissem Energiewerte, welches an den Orten, wo sich Teile des Leiters befinden, einen Strom erzeugt. Ruht aber der Magnet und bewegt sich der Leiter, so entsteht in der Umgebung des Magneten kein elektrisches Feld, dagegen im Leiter eine elektromotorische Kraft, welcher an sich keine Energie entspricht, die aber — Gleichheit der Relativbewegung bei den beiden ins Auge gefaßten Fällen vorausgesetzt — zu elektrischen Strömen von derselben Größe und demselben Verlaufe Veranlassung gibt, wie im ersten Falle die elektrischen Kräfte.

Beispiele ähnlicher Art, sowie die mißlungenen Versuche, eine Bewegung der Erde relativ zum „Lichtmedium" zu konstatieren, führen zu der Vermutung, daß dem Begriffe der absoluten Ruhe nicht nur in der Mechanik, sondern auch in der Elektrodynamik keine Eigenschaften der Erscheinungen entsprechen, sondern daß vielmehr für alle Koordinatensysteme, für welche die mechanischen Gleichungen gelten, auch die gleichen elektrodynamischen und optischen Gesetze gelten, wie dies für die Größen erster Ordnung bereits erwiesen ist. Wir wollen diese Vermutung (deren Inhalt im folgenden „Prinzip der Relativität" genannt werden wird) zur Voraussetzung erheben und außerdem die mit ihm nur scheinbar unverträgliche

34. Einstein introduces special relativity. Above is the beginning of Einstein's celebrated paper of 1905 in which he proposed two postulates: that all inertial (or nonaccelerating) frames of reference are equivalent for the laws of physics and that the speed of light is the same in all such frames even though they are in motion relative to one another. Space and time, that is to say, are relative to the observer. Later, Einstein would deal with accelerating frames of reference in his theory of general relativity, explaining gravity as a curvature of space-time in the presence of mass. The special relativity paper was published in *Annalen der Physik;* an English translation (*right*) did not appear until after Einstein became famous for the predictions of general relativity.

ON THE ELECTRODYNAMICS OF MOVING BODIES

By A. EINSTEIN

IT is known that Maxwell's electrodynamics—as usually understood at the present time—when applied to moving bodies, leads to asymmetries which do not appear to be inherent in the phenomena. Take, for example, the reciprocal electrodynamic action of a magnet and a conductor. The observable phenomenon here depends only on the relative motion of the conductor and the magnet, whereas the customary view draws a sharp distinction between the two cases in which either the one or the other of these bodies is in motion. For if the magnet is in motion and the conductor at rest, there arises in the neighbourhood of the magnet an electric field with a certain definite energy, producing a current at the places where parts of the conductor are situated. But if the magnet is stationary and the conductor in motion, no electric field arises in the neighbourhood of the magnet. In the conductor, however, we find an electromotive force, to which in itself there is no corresponding energy, but which gives rise—assuming equality of relative motion in the two cases discussed—to electric currents of the same path and intensity as those produced by the electric forces in the former case.

Examples of this sort, together with the unsuccessful attempts to discover any motion of the earth relatively to the "light medium," suggest that the phenomena of electrodynamics as well as of mechanics possess no properties corresponding to the idea of absolute rest. They suggest rather that, as has already been shown to the first order of small quantities, the same laws of electrodynamics and optics will be valid for all frames of reference for which the equations of mechanics hold good.* We will raise this conjecture (the purport of which will hereafter be called the "Principle of Relativity") to the status of a postulate, and also introduce another postulate, which is only apparently irreconcilable with the former, namely, that light is always propagated in empty space with a definite velocity c which is independent of the state of motion of the emitting body. These two postulates suffice for the attainment of a simple and consistent theory of the electrodynamics of moving bodies based on Maxwell's theory for stationary bodies. The introduction of a "luminiferous ether" will prove to be superfluous inasmuch as the view here to be developed will not require an "absolutely stationary space" provided with special properties, nor assign a velocity-vector to a point of the empty space in which electromagnetic processes take place.

35. Michelson-Morley experiment.
This sensitive interferometer, mounted on a sandstone slab floating in mercury, was used by the American scientists A. A. Michelson and E. W. Morley in their famous experiment of 1887 that helped to pave the way for relativity theory. Physicists had supposed that light waves in space were carried by an invisible ether, whose existence would lead to interference effects detectable by the apparatus. But the experiment showed that if the ether existed, then the moving earth was at rest with respect to it.

36. Morley-Miller experiment. In 1904, Morley and D. C. Miller built this trussed device, similar to the original apparatus, to test the notion that moving bodies contract in the direction of their motion, a hypothesis that had been advanced to explain the 1887 result. With this interferometer they hoped to detect a difference in contractions. The results were at best inconclusive.

37. The theoretician meets the experimentalists. The three men in the front row of this January 1931 group photo in Pasadena were all winners of the Nobel Prize in physics. A. A. Michelson (*left*) in 1907 became the first American to receive a Nobel Prize in science. Robert A. Millikan (*right*) was the Nobel physics laureate in 1923. As the head of Caltech, he served as host for Einstein and Michelson, who were meeting for the first time.

38. Testing Einstein's theory. General relativity predicted that light from a distant star would be slightly deflected in the sun's gravitational field. The solar eclipse of 1919 presented an early opportunity to test this hypothesis. At left is one of several plates taken by a British expedition at Sobral in Brazil. Short lines mark the stars whose positions were measured. Above is the larger of the Sobral expedition's two astrographic telescopes.

24. Ⅸ. 19

Liebe Mutter!

39. Dear Mother! "Great news today! H. A. Lorentz has telegraphed that the English expedition has really proved the bending of light around the sun." Einstein's postcard is dated September 24, 1919. The eclipse had been on May 29, but analysis of the data, and comparison with existing photographs of the stars' positions when the sun was elsewhere in the sky, had taken some time. The success of the eclipse results catapulted Einstein into international fame overnight.

40. The eclipse telegram. Direct postwar communication between England and Germany was virtually impossible, so news of the British eclipse observations was relayed via Holland to Einstein. The predicted deflection was 1″.7, and the telegram indicates a value between 0″.9 and 1″.8. A more precise value was obtained in 1922 by a Lick Observatory expedition.

41. *Time Transfixed.* René Magritte's surrealistic oil painting from 1938 in-
congruously joins familiar images to evoke a sense of mystery. Although
apparently unaware that the clock and moving train were common meta-
phors from Einstein's explanation of simultaneity and the relativity of time,
Magritte may have been unconsciously sensitized to some of the paradoxes
of relativity theory that were "in the air."

42. The time-space continuum.
The abstract nature of Einstein's theory makes it elusive to illustrate, yet a number of twentieth-century artists have touched on fundamental themes of relativity. The wilting watches in Salvador Dali's 1931 oil painting *The Persistence of Memory* suggest, so to speak, a spatial deformation of time, hinting at the intimate connection between space and time in Einstein's relativity.

43. Elements of measurement. This 1974 etching by the German artist Friedrich Meckseper is part of a series that consciously invokes images for the basic quantities of physics. In *Worpsweden Still Life* these are mass and time, as well as space (in one, two, and three dimensions). The juxtaposition of the clock and cylinder reflects the space-time relation described by Einstein's relativity theory.

SECOND NEWS SECTION

Pittsburgh Post-Gazette

SATURDAY MORNING, DECEMBER 29, 1934.

SPORTS, FINANCIAL, CLASSIFIED SECTION

Constantly Expressive—Einstein Warms Up to His Subject

ATOM ENERG
HOPE IS SPIKE
BY EINSTE

Efforts at Loosing V
Force Is Called
Fruitless.

SAVANT TALKS' HE

Now Indicates Doubt
Relativity Theory H
Made Famous.

44. Einstein meets the press. Soon after taking up permanent residence in the United States, Einstein agreed to participate in the 1934 meeting of the American Association for the Advancement of Science in Pittsburgh. Tickets for his lecture promptly became premium items, and the hall was filled with a crowd that, for the most part, scarcely comprehended his elementary talk on relativity and conservation laws. Einstein had unavoidably become a celebrity, taking on mythical proportions in the mind of the public, and he later tried to avoid such occasions. This account of a Pittsburgh press conference displays fascination with a strange being whose thoughts were not those of ordinary mortals.

45. Einstein, the icon of science. In a 1932 article on the new physics, *Fortune* magazine caricatured Einstein with a wild, almost electric, head of hair. The best-known scientist of his day, the European Einstein had become particularly familiar to the American public because of his winters as a visiting professor at the California Institute of Technology. In 1933, after Hitler became chancellor, Einstein renounced his German citizenship. He eventually accepted a position at the newly founded Institute for Advanced Study in Princeton, New Jersey.

46. Testing the slowing of time. According to Einstein's general theory of relativity, a clock runs more slowly in the vicinity of a massive body. Here the payload is being readied for the Gravitational Redshift Space Probe, which in 1976 carried an extremely accurate hydrogen maser clock high above the gravitating earth. During the experiment, an earth-bound reference clock ran 0.43 billionth of a second slower every second than its spaceborne counterpart at an altitude of more than 5,500 nautical miles (10,200 kilometers)—another confirmation of the general theory of relativity.

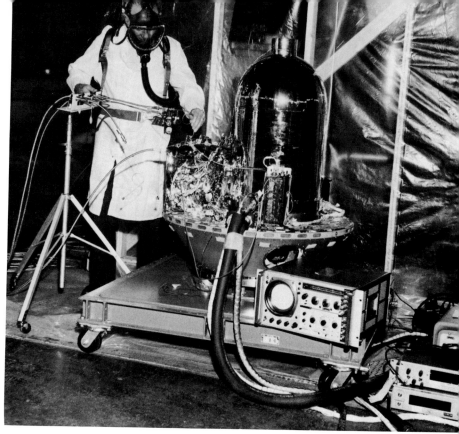

47. Einstein's last page of calculations. These equations from 1955 use tensor algebra, a new form of mathematics that Einstein helped develop and put to physical use in his general theory of relativity. Beginning in 1925 he repeatedly tried to find a "unified field theory," a mathematical formulation that would encompass electrodynamics as well as gravitation, and perhaps ultimately quantum particles as well. Long out of the mainstream of physics, such comprehensive theorizing came into considerable vogue after Einstein's death, as physicists sought to construct so-called grand unified theories, or "GUTs," tying together all the forces of nature.

48. *The Uncertainty Principle.* This 1966 collage by the American artist Joseph Cornell (featuring two images of the nineteenth-century German soprano Henriette Sontag) takes its title from Werner Heisenberg's famous principle, derived from the microscopic world of the quantum. The imagery, however, refers to the paradox of the wave-versus-particle nature of light, epitomized by the interference pattern produced with a double slit, seen in the small central vignette.

CHAPTER

4

Quantum Mechanics

The most powerful and mysterious theory in twentieth-century physics, quantum mechanics originated in the German theorist Max Planck's attempts to find a mathematical formula for the frequency distribution of "blackbody" radiation. This is the electromagnetic radiation emitted by an idealized body that absorbs all radiation falling on it. Blackbody radiation is closely approximated by the radiation emitted from a small hole in a hollow body, and it depends only on the temperature of the body. As the temperature rises, the preponderance of emitted energy shifts to higher and higher frequencies; thus, when a piece of iron is heated, its color changes from red (frequencies mostly at the low end of the visible spectrum) to white (mixture of frequencies throughout the visible spectrum).

In 1900, Planck found a formula that gave an excellent fit to the empirical data on blackbody radiation. He then proposed a mathematical model in which the energy of the system is divided into discrete bundles, later called quanta, that may be distributed in different ways among a number of hypothetical oscillators. It is not clear from his early publications whether he meant that the energy comes in physically discontinuous amounts or is simply treated that way for mathematical convenience in counting the number of possible distributions. In any case he rejected the possibility that electromagnetic energy in space travels in particles rather than waves.

Albert Einstein, in 1905, suggested that one might consider a particle model of light and other

electromagnetic radiation without abandoning the successful wave theory. To explain the photoelectric effect and other phenomena, Einstein proposed that the energy of a quantum of radiation is proportional to its vibration frequency. The proportionality constant h subsequently became known as Planck's constant. Regarding the photoelectric effect, Einstein inferred from his hypothesis that if the electron ejected from a metal has absorbed the energy of a single quantum, its kinetic energy after ejection should increase linearly with the frequency of the incident radiation but should be independent of the radiation's intensity. This prediction was confirmed experimentally in 1916 by Robert A. Millikan.

The quantum hypothesis was applied to atomic structure in 1913 by Niels Bohr. Starting from Rutherford's model of the atom as a tiny, massive, positively charged nucleus surrounded by light, negatively charged electrons moving in a relatively large volume, Bohr used the analogy of the solar system to construct orbits for the electrons. For the hydrogen atom Bohr assumed that there is only one electron, revolving around the proton that constitutes the nucleus. He postulated that only those orbits whose angular momentum is an exact multiple of Planck's constant are allowed, and that the electrons can emit and absorb energy only when jumping from one orbit to another. From these postulates he was able to deduce a previously established formula for the relations between the frequencies of lines in the hydrogen spectrum.

Moreover, he showed that the same model could explain the spectral lines of singly ionized helium atoms, in which there is still only one electron but the nucleus is four times as massive and has two positive charges.

Bohr and others tried to extend his model to many-electron atoms by assuming that the electrons go into the available orbits, starting with those with lowest energy and gradually filling up the others. For the higher energies, elliptical orbits were allowed as well as circular ones. To make this procedure consistent with the known chemical properties of the elements, it was necessary to postulate that each orbit had a maximum capacity of two electrons. In 1925 the Austrian physicist Wolfgang Pauli expressed this limitation as an exclusion principle—each orbit corresponds to two distinct states of the electron, differing in some unspecified additional parameter, and no more than one electron can be in any state. George Uhlenbeck and Samuel Goudsmit in the Netherlands proposed that the additional parameter is related to the rotation, or spin, of the electron—the electron's rotation may be clockwise or counterclockwise.

By 1925 the quantum hypothesis had been generally accepted as a basis for atomic theory. Yet, despite the early successes of Bohr's theory of the hydrogen atom and the prospect that chemical properties could be explained at least qualitatively with the help of the exclusion principle, no one had found a satisfactory way to extend the theory to multielectron atoms and molecules. The interactions between the electrons themselves could not be treated quantitatively without introducing arbitrary additional hypotheses tailored to fit each particular case. The notion that electromagnetic radiation has particle properties had been given further support in 1923 by the American A. H. Compton's analysis of interactions between X rays and electrons—the Compton effect—but it was not clear how these particle properties could be compatible with wave properties such as interference and diffraction. Physicists had not yet found a set of fundamental quantum laws that could replace Newton's laws of mechanics.

The discovery of the quantum laws was made almost simultaneously by two different groups of scientists. In Germany, Werner Heisenberg, working with Max Born and Pascual Jordan, developed a mathematical scheme known as matrix mechanics for describing the absorption and emission of radiation. Meanwhile, in France, Louis de Broglie postulated that electrons have wave properties as well as particle properties, just as electromagnetic radiation has particle properties as well as wave properties. And in Switzerland, Erwin Schrödinger proposed a differential equation to determine these wave properties of the electron and showed that the allowed energy levels of an electron in a hydrogen atom could be derived from this wave equation. Schrödinger's wave mechanics was found to be mathematically equivalent to matrix mechanics but could be much more easily generalized to multielectron systems.

By 1927 matrix mechanics and wave mechanics had been synthesized into a general theory known as quantum mechanics. The wave nature of the electron was experimentally demonstrated by C. J. Davisson and L. H. Germer in the United States and by G. P. Thomson in Scotland. But the physical meaning of the theory—in particular, the interpretation of the wave function appearing in Schrödinger's equation—was still a subject of controversy.

Heisenberg argued that we must give up the assumption, implicit in Newtonian mechanics, that a particle has a definite position and velocity. We may determine its position by an experiment that leaves its velocity unknown, or we may determine its velocity while remaining ignorant of its position. These intrinsic limits on experiments are expressed in Heisenberg's uncertainty principle. Max Born proposed that the wave function tells us only the *probability* that a measurement of the electron will give a particular result.

Einstein objected to the implication, apparently essential to quantum mechanics, that the behavior of atomic particles has an element of inherent randomness; "God does not play dice," he asserted. But Niels Bohr, in a series of discussions with Einstein, argued that quantum mechanics nevertheless gives a complete and consistent description of the observable properties of atoms, even though this leads to some remarkable paradoxes if one insists that a physical system really has a definite physical state when it is not being observed. For example, Schrödinger's "cat paradox" imagined a cat sealed in a box with a random device that had a 50 percent chance of killing it. According to quantum mechanics, the cat is neither dead nor alive until someone unseals the box to observe it. Thus, quantum mechanics almost seems to deny that the world exists apart from human observation of it.

Stephen G. Brush

49. Birth of quantum physics.

Quantum theory, one of the most fundamental and far-reaching revolutions of twentieth-century science, was born with the century. In 1900 the German theoretical physicist Max Planck introduced the hypothesis that the frequency distribution of radiant energy was the same as it would be if the energy could be emitted or absorbed only in multiples of a basic unit, or quantum, whose size was proportional to the radiation's frequency. The proportionality constant h became known as Planck's constant. Reproduced here are extracts from Planck's paper, published in 1901 in *Annalen der Physik*. Conservative by nature, Planck was reluctant to view the quantum hypothesis as more than a useful fiction. The constant k, it might be noted, was given its name by Planck: the Boltzmann constant. It plays an important role in thermodynamics.

50. Blackbody emission curves.

Classical physics could not account for the distribution of energy in radiation emitted by a blackbody, an ideal body that absorbs all radiation incident on it, reflecting none. However, from Planck's quantum hypothesis there followed a formula that did account for the experimental data. The graph at right, from a 1901 paper, compares the observed data for constant frequency but variable temperature with the emission curves predicted by four scientists (Wien, Thiesen, Lord Rayleigh, and Planck). Planck's curve best fits the data.

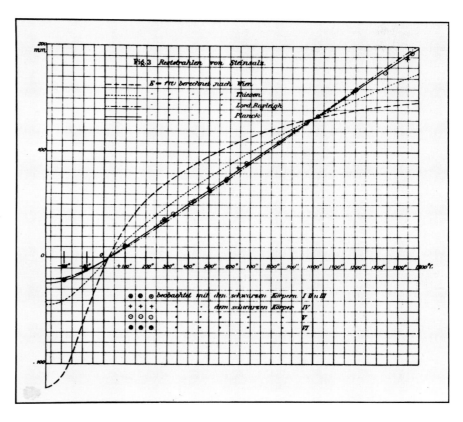

51. Early X-ray diffraction equipment. For over a decade following Röntgen's discovery of X rays, scientists debated their nature. In 1912 the German physicist Max von Laue used the diffraction pattern from regular crystal structures to prove that X rays were electromagnetic radiation of short wavelength. The technique could be used in reverse, to explore the structure of materials. Here we see, in a view dating from about 1923, the apparatus at the Fiber Research Institute of the Kaiser Wilhelm Institute in Berlin.

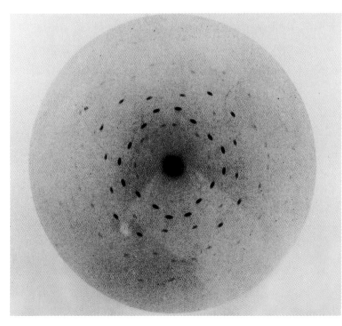

52. X-ray diffraction. This 1912 image is the first photograph of the diffraction of X rays by zinc sulfide. The pattern of black spots indicates that zinc sulfide has a cubic crystal structure. Von Laue's younger colleagues Walter Friedrich and Paul Knipping obtained the photograph after von Laue suggested that if crystals are regular structures of atoms, and if X rays are electromagnetic waves of very short wavelength, then diffraction images might be found. A decade later, a similar technique allowed the American physicists Clinton Davisson and Lester Germer to demonstrate the wave nature of electrons.

53. Bohr's atomic theory. According to the theory propounded by Niels Bohr in 1913, when an atom emits light, the light's frequency ν is proportional to the difference between the atom's energy states before and after emission.

54. Energy levels for Bohr's hydrogen atom. In Bohr's theory, atoms could emit (or absorb) radiation only in discrete, or "quantized," amounts as their electrons jumped from one allowed orbit to another. The diagram shows the specific energy levels of the hydrogen atom's orbits according to Bohr's model; the lines connecting the levels in the diagram correspond to the permitted radiation quanta, and thus to the actual series of lines in the spectrum of hydrogen. The left-hand scale shows the energy, in electron volts.

55. Bohr's notes, 1921. The 1913 theory arranged electrons in definite, concentric rings. Bohr revised his model, so that the electrons traveled in elliptic orbits. The new theory yielded definite structures for all the atoms of the periodic table. Although incorrect in some details and soon superseded, the model provided the first theoretically grounded scheme for all the elements.

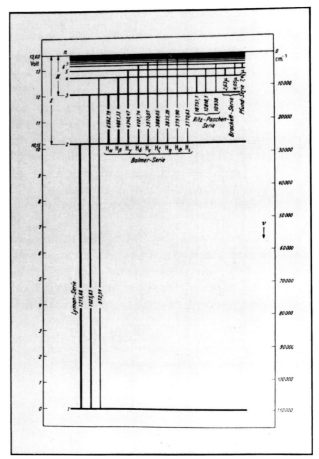

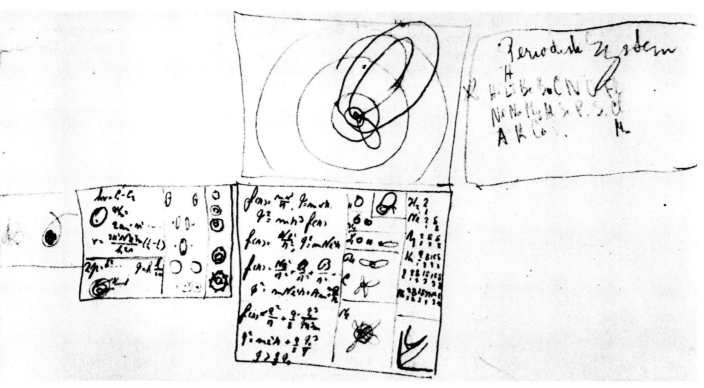

56. Zeeman, Einstein, and Ehrenfest. The Dutch physicist Pieter Zeeman (*left*) discovered in 1896 the Zeeman effect, which was to play a key role in decoding atomic structure. Einstein, though best known for relativity theory, made seminal contributions to quantum mechanics. The Austrian Paul Ehrenfest, who succeeded H. A. Lorentz as professor of theoretical physics in Leiden, was noted for his critical analysis of quantum mechanics, as well as his personal enthusiasm for science. He was responsible for several historic discussions between Einstein and Bohr, who were close friends of his. The photograph was taken in Zeeman's laboratory in Amsterdam.

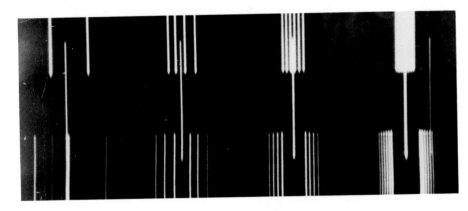

57. Zeeman effect. When a source of light is placed in a strong magnetic field, the spectral lines split into several components. The Zeeman splitting results from the interaction of the magnetic field with the motion of electrons in the atom. The vanadium spectrum at left was made at the Massachusetts Institute of Technology in the late 1940s. The upper spectra are plane polarized; the lower ones, circularly.

58. Proof of space quantization. Quantum theory required that neutral atoms be able to orient themselves in just a few discrete (quantized) directions in a magnetic field. In 1921 the German physicists Otto Stern and Walther Gerlach shot a narrow beam of atoms (which, according to Bohr's quantum mechanics, should have only two orientations) through a nonuniform magnetic field and onto a glass plate. The beam split into two parts, as photographed here, just as the theory predicted.

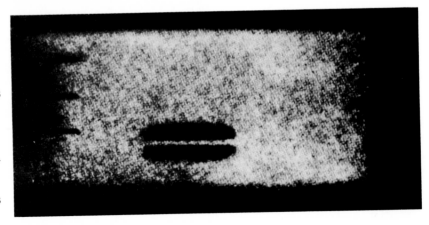

59. Heisenberg and Bohr. The German physicist Werner Heisenberg (*left*) put forth his uncertainty, or indeterminacy, principle in 1927. The principle states that it is impossible to determine accurately at the same time both the position and the momentum of a particle. The photograph was taken at Bohr's Copenhagen institute in 1934. Intellectual camaraderie and lively discussion were the order of the day at the institute, as here between Bohr and Heisenberg.

60. Heisenberg's uncertainty. The notion of indeterminacy even provided material for this bumper sticker from the American Chemical Society. But the principle has a specific meaning in quantum mechanics, where experimental measurement of one observable variable produces uncertainties about the value of others.

Heisenberg may have slept here

61. Fifth Solvay Conference: quantum mechanics comes of age. The Solvay conferences, established by the Belgian industrial chemist Ernest Solvay, brought distinguished physicists to Brussels beginning in 1911. In 1927 the conclave considered quantum mechanics. Seated in the front row (*left to right*): Langmuir, Planck, Curie, Lorentz, Einstein, Langevin, Guye, Wilson, and Richardson; in the second row: Debye, Knudsen, Bragg, Kramers, Dirac, Compton, de Broglie, Born, Bohr; standing in back: Piccard, Henriot, Ehrenfest, Herzen, De Donder, Schrödinger, Verschaffelt, Pauli, Heisenberg, Fowler, Brillouin.

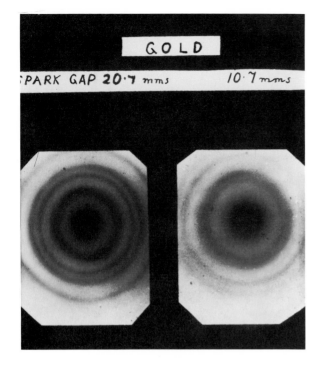

62. Electrons as waves. In the early 1920s, the French physicist Louis Victor de Broglie suggested that, just as light had been found to have both wave and particle properties, matter (such as electrons) might have wave in addition to particle properties. This was indeed so: just as light waves passing through a pinhole could produce characteristic diffraction rings, so could electrons shot through gold foil, as recorded here in 1927 by George P. Thomson, the son of the discoverer of the electron.

63. Demonstrating the wave nature of matter. Clinton Davisson and Lester Germer of Bell Laboratories demonstrated electron diffraction in 1927 by reflection from a crystal of nickel. The distribution of reflected, or scattered, particles was just as if X rays had been used—the electrons were behaving like waves. The picture shows Germer seated at the observer's desk, apparently about to read the galvanometer, which is visible next to his head. Behind Davisson (*left*) are the batteries that powered the experiment. At right is the scientists' assistant.

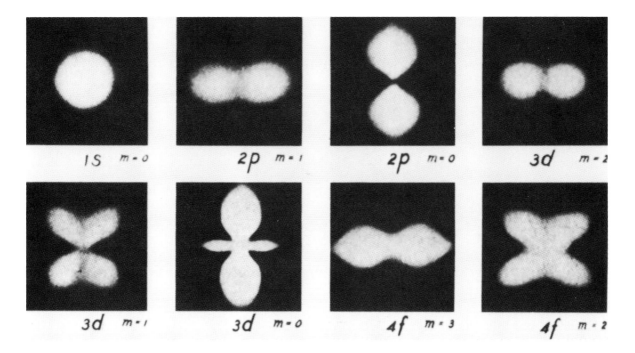

64. Schrödinger's electron clouds. In 1926 the Austrian-born physicist Erwin Schrödinger developed a fundamental equation describing the behavior of electrons as waves. Where Bohr envisioned electrons in elliptical orbits around the nucleus, Schrödinger depicted the atomic structure as diffuse clouds showing where it was most probable to find an electron. (This indeterminacy in locating the electron is related to Heisenberg's uncertainty principle.) These theoretically derived pictures show the probability patterns for some of the states of the hydrogen atom. Each image is characterized by its "quantum numbers."

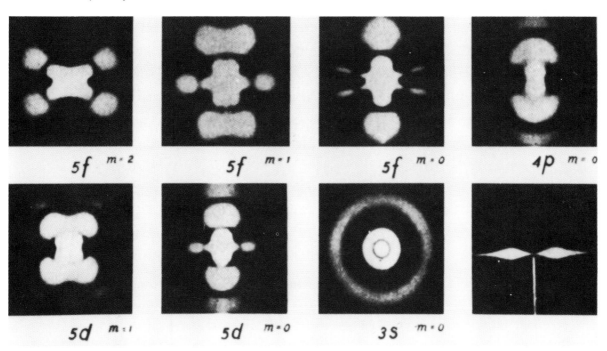

65. Summer in Ann Arbor, 1930.
Beginning in the late 1920s the summer school of the University of Michigan brought leading physicists to America. Such occasions nurtured the first flowering of theoretical physics in the United States. At far left is German-born Maria Goeppert Mayer, who would later receive the Nobel Prize for helping to develop the shell theory of the atomic nucleus. Next to her is her husband, the American chemist Joseph Mayer. The others are the English astronomer Robert d'E. Atkinson, Paul Ehrenfest, and the Norwegian-born chemist Lars Onsager.

66. 1930 Copenhagen conference. Bohr's Institute for Theoretical Physics in Copenhagen opened in 1921 and promptly became a major center for research in atomic structure, drawing a steady stream of visiting physicists. Annual conferences were held beginning in 1929. Sitting in the front row in this photograph are (*left to right*) the Swedish physicist Oskar Klein, Bohr, Heisenberg, the Austrian Wolfgang Pauli, the Russians George Gamow and Lev Landau, and the Dutchman Hendrik Kramers.

67. Cesium-beam clock. Practical applications of quantum theory include the atomic oscillator clock, which uses transitions between atoms' energy levels to provide a frequency standard. Cesium atoms of isotope 133 proved to be particularly effective, and in 1967 the second was defined in terms of cesium 133's natural resonance: 9,192,631,770 cycles per second. When used as a clock, this cesium-beam apparatus at the U.S. National Bureau of Standards, could keep time with an accuracy of three microseconds a year.

68. Planck's constant and the new and old physics.
Sometimes it's said that classical physics is what you would get if the speed of light were infinite and Planck's constant h zero: no relativistic effects, and no quantum discreteness imposed by a nonzero h. But this way of thinking is misleading. One day in 1961, when such matters were being discussed, Niels Bohr wrote on the blackboard the so-called fine-structure constant: the square of the electron charge e divided by the product of Planck's constant and the speed of light c. This fundamental constant yokes together elements of electron theory (e), quantum theory (h), and relativity theory (c). Bohr underlined the h and said, "You see, h is in the denominator." Setting h equal to zero would create a meaningless quantity involving division by zero.

69. Holography. The principles of holography were developed in 1947, but the invention of the laser in the early 1960s made it a practical reality. The laser exploits a quantum effect, first described by Einstein in 1917, to trigger a powerful "coherent" beam, that is, one in which all the light waves are in step. Holography records interference patterns from the reflected laser light, using both the intensity and the light's phase.

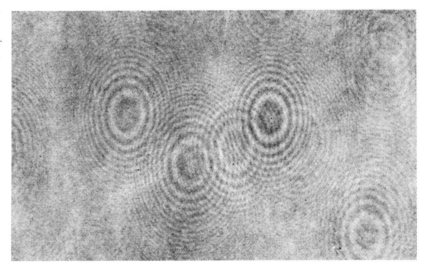

The three pictures on this page are from holography's infancy in the early 1960s. They indicate how a three-dimensional image is produced. At the top is an ordinary photograph of a chessboard. The middle picture is a hologram, recording the same scene. The bottom picture is a three-dimensional image of the original scene, made with a laser beam and the hologram.

70. Making a hologram. This photograph shows how a laser produced the holographic interference pattern on the opposite page. The beam enters at top right. Two glass plates partially reflect and partially transmit the laser light. Two mirrors (*bottom*) reflect the laser light onto the chessboard. On the hologram plate (object near the chessboard at bottom center), light reflected from the chess pieces interferes with the reference beam (reflected from upper left) to produce the hologram.

Part Two

THE STARS
AND BEYOND

71. Mount Wilson's 100-inch reflector in action. Astrophysicist-entrepreneur George Ellery Hale got money from businessmen Andrew Carnegie and John Hooker to build what was then the world's largest telescope. The 100-inch (2½-meter) instrument was completed in 1917. Here Francis Pease, instrument designer as well as astronomer, sits at the eyepiece. Pease helped mount a huge interferometer on the telescope in 1920 in order to measure the size of the red supergiant star Betelgeuse.

5

Building Large Telescopes

Like buried treasures, the outposts of the universe have beckoned to the adventurous from immemorial times. Princes and potentates, political or industrial, equally with men of science, have felt the lure of the uncharted seas of space, and through their provision of instrumental means the sphere of exploration has rapidly widened. If the cost of gathering celestial treasure exceeds that of searching for the buried chests of a Morgan or a Flint, the expectation of rich return is surely greater and the route not less attractive. . . . The latest explorers have worked beyond the boundaries of the Milky Way in the realm of the spiral "island universes," the first of which lies a million light-years from the earth. . . . While much progress has been made, the greatest possibilities still lie in the future.

This stirring call to celestial exploration appeared in the April 1928 issue of *Harper's Magazine*. Its author, the American solar astronomer George Ellery Hale, made sure that a copy came to the desk of Wickliffe Rose, an influential member of the Rockefeller Foundation. The article proved to be a powerful catalyst in raising the $6 million required to build the 200-inch, or 5-meter, reflector on Palomar Mountain, an instrument that finally saw "first light" in 1948.

The incident illustrates elements that helped place the United States in the forefront of telescope building in the century between 1880 and 1980. In the first place, the clear skies and steady seeing on mountain tops in the American West offered the most favorable sites for telescopes within the industrial countries able to support astronomy. Secondly, a freewheeling capitalist economy produced a series of millionaires eager to polish their images through public philanthropy. Both factors first came together in 1876 when the San Francisco real estate magnate James Lick was persuaded to endow an unprecedentedly large telescope on Mount Hamilton in the Diablo Range southeast of the Bay Area. The success of the Lick Observatory's 36-inch refractor paved the way for Hale's brilliant achievements as the greatest scientific entrepreneur of the twentieth century.

The first striking demonstration of Hale's characteristic verve came in 1892, when he was still a young astronomer working in Chicago. On learning, during a meeting of the American Association for the Advancement of Science, that the glass for a 40-inch lens might be available, he packed up, left the meeting, and headed straight back to Chicago to find a philanthropist to fund a 40-inch refracting telescope. Hale convinced the wealthy trolley financier Charles T. Yerkes to provide $349,000 for the world's largest refractor for the University of Chicago, along with an observatory to bear Yerkes's name. Built not far from Chicago in southern Wisconsin, the new facility gave Hale the opportunity to study the sun by day while staff examined stars and their spectra by night.

In quest of even better skies, Hale migrated to southern California, and by 1904 he had founded a solar observatory on Mount Wilson, overlooking the Los Angeles basin, with funds from the Carnegie Institution of Washington, a private organization founded by Andrew Carnegie for the advancement of scientific research. Hale started with a horizontal telescope that gave a solar image 16 centimeters (about 6 inches) wide, but he was keen to extend his spectrographic analyses to the stars as well, and for this he needed far more light-gathering power. James Keeler's work with a 36-inch reflector at Lick Observatory had strikingly demonstrated the efficacy of reflecting telescopes. A reflector required the polishing of only a single optical surface, and its primary optical element could be supported from the back, thereby accommodating a much wider but not appreciably thicker glass than the refractor at Yerkes. A 60-inch glass blank had become available in France, and Hale's father purchased it for him. Heroic work was required to cut a trail up the mountain so that heavy metal pieces could be carried to the summit. With further funds from the Carnegie Institution, Hale completed the world's largest telescope on Mount Wilson by 1908, and two years later Carnegie himself came to have a look at it.

Ever enthusiastic to penetrate deeper into the skies, Hale undertook to extend his observatory with a 100-inch reflector as well as a 150-foot solar tower. John D. Hooker, a Los Angeles businessman, agreed to pay for the $45,000 glass mirror for the big reflector, and the Carnegie Institution eventually furnished the $600,000 required for its installation. Completed in 1917, the 100-inch giant soon proved its merit. By 1918, Harlow Shapley, working with both the 60-inch and 100-inch, showed that the Milky Way was far more extensive than previously envisioned, and a few years later Edwin Hubble demonstrated that the "island universes" lay a million light-years away and beyond (see Chapter 6). Meanwhile, Hale achieved his goal of accompanying the telescopes with a fully equipped spectroscopy laboratory that probed the nature of atomic spectra, vital for analyzing the composition and structure of the sun and stars.

With the 100-inch bringing in wonderful new results on the expansion of the universe, Hale dreamed of an even larger eye on the universe. "I believe that a 200-inch or even a 300-inch telescope could now be built and used to the great advantage of astronomy," he wrote in his *Harper's* article. This project proved far more difficult than Hale's previous record-breaking telescopes. Early attempts to provide a low-expansion glass failed, but a suitable Pyrex disk was finally cast by Corning Glass in New York State in 1934. Hale himself did not live to see his vision fulfilled, dying in 1938, a decade before the glass giant of Palomar was finally installed.

The success of Hale's reflectors and their technological solutions encouraged other observatories to think big. The Dominion Astrophysical Observatory in Victoria, British Columbia, acquired a 72-inch reflector in 1918. And Shapley (who had left Mount Wilson for Harvard) prevailed on the Rockefeller Foundation to establish a 60-inch in South Africa—at the time, the largest telescope in the southern hemisphere. In 1925, William J. McDonald, a Texas banker, unexpectedly bequeathed nearly a million dollars to the state university for an observatory bearing his name. Opened in 1939, the 82-inch reflector on Mount Locke in western Texas was number two after the Mount Wilson 100-inch.

European observatories, lacking advantageous climates, fell behind in their observing facilities. Eventually, with increasingly enhanced travel possibilities, European astronomers took advantage of the favorable climates in Chile and Hawaii to compete with the American observatories (see Chapter 22).

Owen Gingerich

72. Yerkes 40-inch refractor. At the turn of the century, the biggest refractor and one of the largest telescopes of any kind was the 40-inch (1-meter) instrument at Yerkes Observatory in Williams Bay, Wisconsin. At right is the mounting on display at the 1893 World's Columbian Exposition in Chicago.

73. Building the dome for the 40-inch. George Ellery Hale, then at the University of Chicago, was the driving force behind construction of the telescope and observatory. He raised the funds from trolley-car magnate Charles T. Yerkes, who wished to refurbish his reputation. The observatory was dedicated in 1897. Its refractor remains unrivaled in size and continues today as an instrument for astronomical research.

74. Master opticians of the 40-inch lens. The Cambridge, Massachusetts, firm of Alvan Clark & Sons made the lenses for many of the major telescopes of the nineteenth century, repeatedly breaking the world record. Alvan G. Clark (the last surviving optician of the family) and Carl A. R. Lundin (*right*) worked for five years to prepare this high-water mark of lens making, finally finishing the job in 1895.

75. Up Mount Wilson by burro.
When Hale visited southern California in 1903, he immediately recognized the advantages of the western climate compared to Wisconsin's. In 1904 he founded the Mount Wilson Solar Observatory in the mountains overlooking Pasadena. In the beginning the site was isolated, and conditions were primitive. Supplies were carried to the top by burro train, which took an entire day to negotiate the narrow trail.

76. Carnegie and Hale. In order to equip his new observatory with world-class instruments, Hale once again pursued America's leading philanthropists. The Carnegie Institution approved building a 60-inch (1½-meter) reflector, and the telescope was set up in 1908. In March 1910, Andrew Carnegie himself paid a visit to Mount Wilson. In the photograph we see him (*left*) and Hale beside the 60-inch telescope. Until the completion of the 100-inch Hooker reflector, the 60-inch was the world's largest telescope in effective use.

77. An accident on the trail. In the initial stages of construction on Mount Wilson access to the observatory site remained difficult, but in 1906–1907 the trail was widened to allow for carriages. A truck took up the structural steel for the 60-inch telescope and subsequently the larger parts for the 100-inch. This photograph from about 1916 catches a moment when a precipice almost claimed the central tube section for the 100-inch and the truck hauling it.

78. Mount Wilson from the air. This view from the southeast shows on the left the 60-foot (18-meter) and 150-foot vertical tower telescopes used for solar observations. The two large domed structures contain the 60-inch and 100-inch reflectors.

79. Hale's 200-inch telescope at Palomar. After Hale retired as director of Mount Wilson, he launched a campaign to build a 200-inch reflector. In 1928 the Rockefeller Foundation granted $6 million to the California Institute of Technology for the telescope. Among its many innovations was a modified yoke mount, which allowed the giant instrument to reach the north pole region of the sky; the design was widely copied in later reflectors. This 1939 drawing is part of a memorable series by the artist and amateur telescope maker Russell W. Porter.

THE · TWO · HVNDRED · INCH ~
TELESCOPE · LOOKING · NORTHWEST

80. Making the 200-inch mirror. Attempts to make a quartz mirror were unsuccessful, and the Corning Glass Works in New York was contracted to produce a Pyrex disk. In 1934 the Pyrex glass was poured into its mold (*above left*). After months of annealing the 200-inch disk was shipped to California for finishing. The lower photograph shows the 160-ton grinding and polishing machine. The tools were attached to the movable bridge at the top of the machine. The turntable on which the mirror lies could be tilted to put the mirror in a vertical position for testing.

81. Hubble in the observer's cage. The 200-inch telescope was so large that the observer could sit in a cage at the prime focus, rather than having the focused light beam reflected to the side of the tube or back through a hole in the mirror as in previous reflectors. Here Edwin Hubble, the American astronomer who established the great distances to the galaxies and the observational basis for the expanding universe, takes a turn in the giant Hale telescope. Reflected in the mirror at lower left are portions of the telescope's structure and dome.

82. Schmidt in his optical shop. In 1931 the German optical instrument maker Bernhard Schmidt devised a telescope that combined a spherical mirror with a thin correcting lens. The ordinary parabolic reflecting mirror, such as the Palomar 200-inch, had a limited field of view because optical aberrations off the optical axis blurred the images. The Schmidt telescope's excellent image definition and large field of view made it well suited for photographing broad sections of sky. Schmidt polished his mirrors by hand, working alone although he had lost his right arm in an accident in his youth. The photograph was taken at the Hamburg Observatory, Bergedorf, in 1928.

83. Schmidt telescope at Palomar.
This Russell Porter drawing depicts
the Schmidt instrument that went
into operation on Palomar Mountain in 1948. At the upper end of
the tube is the glass correcting
plate, which compensates for the
image defects known as spherical
aberration and coma; it is 48 inches
(1.2 meters) in diameter. At the
lower end is the mirror, 72 inches
across. For more than a decade this
was the world's largest Schmidt
telescope.

84. 82-inch McDonald telescope.
After its completion in 1939, this
reflecting telescope on Mount
Locke, Texas, remained for nearly
a decade the second largest in the
world. The project was partly en-
dowed by the Texas banker William
J. McDonald. The instrument was
initially operated in collaboration
with astronomers from Yerkes Observatory. Its early uses included
not only stellar spectroscopy but
also solar system studies, which was
unusual for such a large telescope.
Among the discoveries made with
this instrument were Miranda (a
moon of Uranus) and Nereid (a satellite of Neptune).

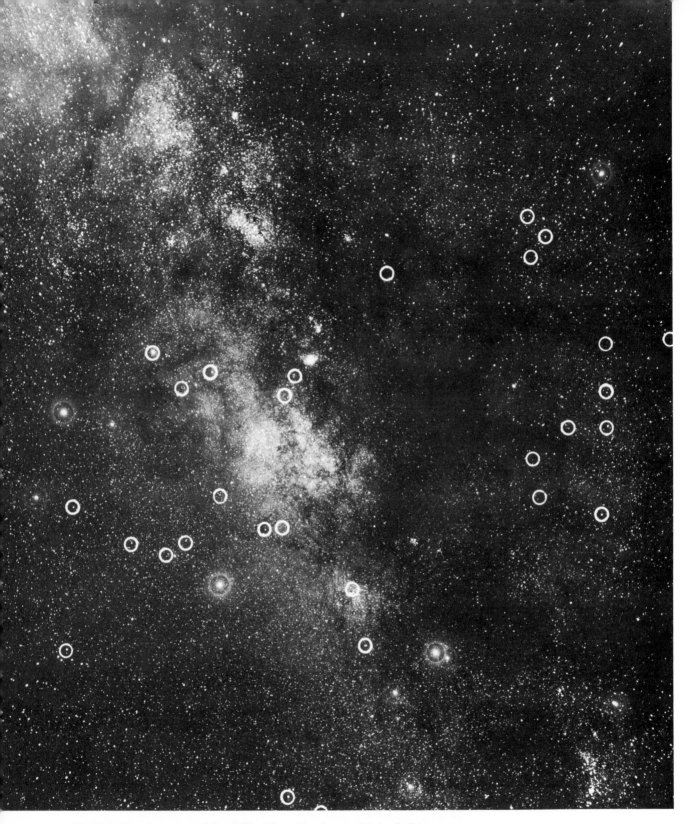

85. Globular clusters and the Milky Way. Nearly one-third of all known globular clusters in our galactic system are found in this small region of the constellation Sagittarius, about 3 percent of the entire sky. Working from this fact and from his distance determination to the globulars (here marked with circles), the American astronomer Harlow Shapley in 1918 surmised that our Milky Way was a much vaster star system than previously believed.

6

The Scale
of the Universe

Let the circumference of a dime represent the annual orbit of the earth around the sun. How large would the universe of stars and nebulae available to the naked eye be on this scale model?

At the turn of the century, astronomers had long believed the stars to be like the sun. If the sun were moved far into space, it would remain visible until its distance exceeded 50 light-years or so. Thus, it was thought that the faintest stars visible to the naked eye, those of sixth magnitude, would lie 50 light-years away—a distance in the dime model of about 20 miles, or 30 kilometers.

By 1920 astronomers realized that some stars were intrinsically far brighter than the sun. A few researchers were convinced that the most luminous Cepheid variables—stars that rhythmically varied in their light intensity—were 10,000 times as brilliant as the sun. Hence, the faintest naked-eye Cepheid variable could be a far beacon 5,000 light-years away, and the dime model would have a radius of 2,000 miles. This conclusion was controversial.

At Harvard, Henrietta Leavitt had shown that the brightest Cepheids in the Small Magellanic Cloud had the longest periods, and this led Harlow Shapley, then at George Ellery Hale's Mount Wilson Observatory, to infer that the brightness was intrinsic to their structure. In other words, all Cepheids with periods of ten days would have the same intrinsic brightness no matter where they were found, but their apparent brightness would, of course, depend on how far away they were.

Acting on this hunch, Shapley used the apparent brightnesses of Cepheids found with the aid of the telescope in globular star clusters to estimate the distances to the clusters. He calculated that the great globular cluster M13 in the constellation Hercules—near the limit of naked-eye visibility—was 30,000 light-years distant. The model with the dime for the earth's orbit would then place M13 about 12,000 miles away. Furthermore, Shapley noticed that the overwhelming majority of the globulars were concentrated in only half the sky, and so he concluded that the center of our Milky Way system lay at a great distance from the solar system, in the direction of the constellation Sagittarius. Many astronomers were reluctant to accept a universe so huge and so eccentric.

Meanwhile, 300 miles to the north, at the rival Lick Observatory, Heber D. Curtis was studying the "new stars," or novae, that he found in a handful of faint spiral nebulae. The novae seemed to be exploding stars of unusual luminosity, intrinsically brighter than even the Cepheids. His rough calibration of their intrinsic luminosity led to a distance of about a million light-years for the nebula M31 in Andromeda. A spiral whose intense nucleus was easily visible to the naked eye, M31 would in the dime model be more distant than the moon, and fainter spirals would lie still farther away!

Shapley, for one, found Curtis's distances to the spirals incompatible with his views on the immense size of the Milky Way system. M31 seemed far smaller than the dimensions he obtained for our own galactic system, and he placed the spirals within the immediate neighborhood of the Milky Way. Curtis, on the other hand, not only thought that the the spirals were distant island universes but also firmly believed that Shapley had greatly overestimated the luminosity of the Cepheid variable stars, and hence the size of the Milky Way. The stage was set for a historic debate before the U.S. National Academy of Sciences in April 1920. The verdict of passing years is that each man turned out to be partially right and partially wrong, but the enlarged scale of the universe, which each in his own way supported, remains as an impressive legacy.

Shapley soon became director of the Harvard College Observatory, and Curtis director of the Allegheny Observatory in Pittsburgh. Back at Mount Wilson, Edwin Hubble carried on researches into the nature of the nebulae, and early in 1924 he wrote to Shapley confiding his discovery of Cepheid variables in the Andromeda nebula. This finding confirmed Curtis's large distances for the spirals. A few years passed before other evidence showed that Shapley's picture of the Milky Way was in principle correct. In Europe the Swedish astronomer Bertil Lindblad and the Dutchman Jan Oort discovered that the Milky Way rotated about a distant center located in the direction of the constellation Sagittarius, a finding that corroborated Shapley's view of a vast, pancake-shaped galaxy with the solar system lying far from its nucleus.

Lingering questions remained about the apparent disparity in size between the Milky Way and the great M31 spiral in Andromeda. In 1930, Robert Trumpler's discovery of interstellar absorption (see Chapter 7) showed that the Milky Way is a foggy system in which remote stars appear faint because of obscuration as well as distance; this meant that some of their distances had been overestimated. The sun still lay off-center in the Milky Way, whose dimensions still dwarfed those of M31, but the discrepancy was less.

Meanwhile, Edwin Hubble continued his investigations of the spirals. He knew, from the work of Vesto M. Slipher at the Lowell Observatory in Arizona, that some of these nebulae exhibited spectra greatly shifted with respect to laboratory measurements, and this suggested that they had extraordinarily high velocities compared with even the fastest-moving stars. By 1929, using Mount Wilson's great 100-inch and 60-inch reflectors, Hubble found an astonishing correlation: the fainter the nebulae, the more their spectra were shifted toward the red wavelengths. Assuming that the fainter nebulae were the more distant, and that the red shift represented a Doppler velocity of recession, then the more distant a nebula was, the faster it was rushing away from us. This finding soon led to the idea of the expanding universe, perhaps the most remarkable of all twentieth-century astronomical conceptions.

While Hubble was gathering his observational data, theoreticians had begun to explore some of the ramifications of Einstein's general theory of relativity, which was in effect a theory of gravitation and presumably relevant to the combined mass of the universe as well as to individual objects. By the late 1920s a few mathematical physicists realized that Einstein's equations would allow for a dynamic universe that was expanding or contracting, and the astronomers' observations quickly settled which was the case.

The expansion implied an earlier, denser state of the universe. It was even possible to calculate the time when the entire universe was "crushed in one harbor," to borrow poet Robinson Jeffers's evocative phrase. Surprisingly, the age of the universe came out to only about 2 billion years, which was considerably shorter than geologists felt was a sufficiently long span for the evolution of the earth. The conflict remained unresolved until 1952, when Walter Baade at Mount Wilson and Palomar observatories provided a wholesale recalibration of the Cepheid distance scale. This not only produced a larger and older universe but also cleared up the puzzle as to why our own Milky Way seemed to be the giant among the spirals. When placed at a greater distance, the Andromeda spiral indeed turned out to be the same size as, or even a little bigger than, our own galaxy. The same year brought confirmation of what astronomers had for some decades suspected and even believed, that the Milky Way was in fact a spiral pinwheel (see Chapter 19). Thus, the first half of the twentieth century saw astronomers delineate the size and nature of the Milky Way and recognize its place within a vast array of galaxies rushing away from each other with astounding speeds.

Owen Gingerich

86. Small Magellanic Cloud. A satellite companion to our Milky Way galaxy, the SMC is faintly visible to the naked eye in the southern skies. This five-hour exposure was made with the 24-inch (61-centimeter) Bruce doublet refractor at the Boyden Station of Harvard College Observatory at Arequipa, Peru, in 1898. It was one of the plates used to establish the period-luminosity relation of Cepheid variables.

87. Period-luminosity relation. When Harvard's Henrietta Leavitt determined the cycles of light variation for twenty-five Cepheid variable stars in the Small Magellanic Cloud, she found that the brightest stars had the longest periods. This graph, published in 1912, shows a direct relation between a star's apparent magnitude and the logarithm of its period. After Harlow Shapley calibrated their absolute magnitudes, the Cepheids became the basis for a powerful distance-measuring method.

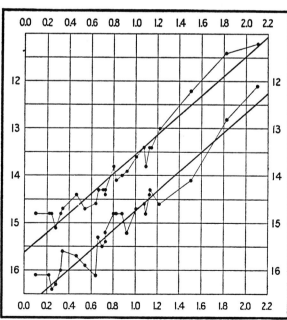

88. Novae far, far away. In 1917, Heber D. Curtis discovered two novae in the spiral nebula NGC 4321 while studying the above two plates, taken at the Lick Observatory on Mount Hamilton in California. The left one was made in 1901, and the right one in 1914. At about the same time as Curtis's discovery, other novae were also being found in photographs of spiral nebulae. The faintness of such novae, when compared with those in the Milky Way, suggested that the spiral nebulae were remote island universes far beyond our own galactic system.

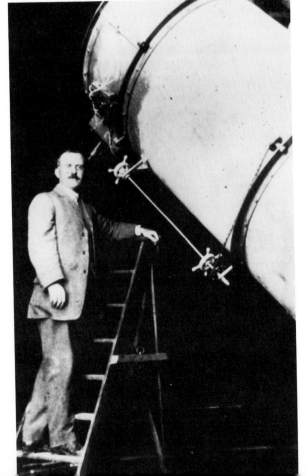

89. Curtis at the Crossley reflector. This 36-inch (91-centimeter) reflector was considered a white elephant by the Lick Observatory staff until James Keeler demonstrated that it could take superlative photographs of nebulae. Curtis, his successor as Lick's nebular specialist, became an outspoken advocate of the notion that the spirals are galaxies far outside the Milky Way.

90. An island universe, M81. George W. Ritchey, the optician who oversaw the polishing and figuring of the 60-inch and 100-inch reflectors at Mount Wilson, used part of his allotted time at the telescopes to take a series of "show plates" of some of the most spectacular deep-sky objects. He photographed the spiral nebula M81, in the constellation Ursa Major, with the 60-inch telescope around 1912.

91. Nebulous region in Virgo. This wide-angle photograph shows the central part of what is now recognized as a large cluster of galaxies in the constellation Virgo. Edward E. Barnard, who specialized in wide-angle views of the Milky Way and nebulae made possible with a comparatively small, short-focus instrument, took this picture with the 10-inch (25-centimeter) Bruce telescope, possibly in 1905 when it was sent to Mount Wilson.

92. Globular star cluster, M22. This cluster, one of the nearer globulars in Sagittarius, was the first recorded by astronomers, in 1665. The system contains approximately 70,000 stars. This three-and-a-half-hour exposure was made with the Mount Wilson 60-inch reflector in 1918.

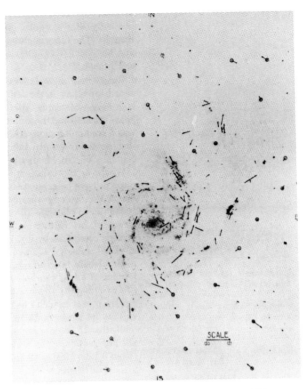

93. Internal motions in the spiral nebula M101. The Dutch-born astronomer Adriaan van Maanen published in 1916 an analysis of apparent rotation in M101. The arrows on this Mount Wilson plate, from van Maanen's paper, show the relative magnitude and direction of motion for the points he measured. Shapley argued that if the spiral was far away, the observed motion would mean impossibly fast rotational velocities. However, van Maanen's findings for M101 had to be wrong; they were never confirmed by anyone else.

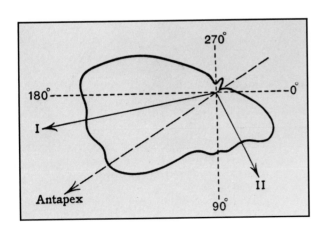

94. Star streaming. In 1904 the Dutch astronomer J. C. Kapteyn noticed that in each part of the sky stars tended to drift in two preferential directions. Subsequently the British astronomer Arthur Eddington discerned that when the sun's own motion was subtracted out, all stars tended to move in the same two streams; this diagram for one region is from Eddington's 1914 book. After Shapley proposed his galaxy model, it was discovered that star streaming was a natural consequence of galactic rotation.

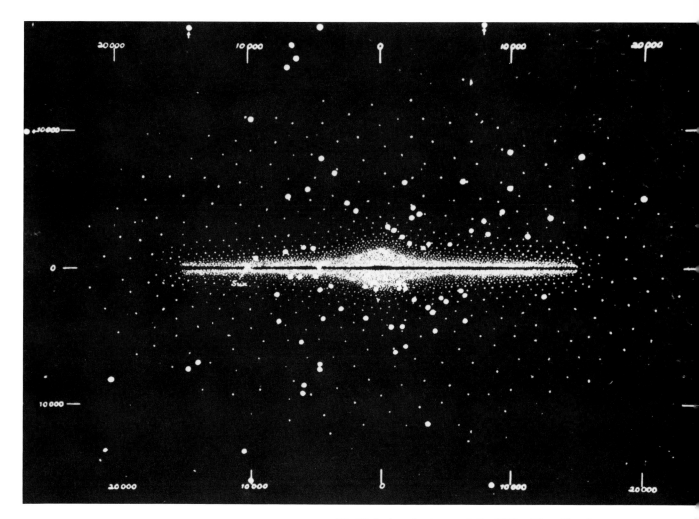

95. Globular clusters and the Milky Way. Shapley imagined the Milky Way as a disklike system (here shown edge-on) with a nucleus centered among the globular clusters. He originally determined that the center was about 50,000 light-years from the sun. As this diagram made by J. S. Plaskett in 1935 shows, the figure was later reduced, to about 10,000 parsecs (around 30,000 light-years). The dark rift dividing the galaxy represents a belt of interstellar "fog."

96. Organizing the Great Debate. At the end of the second decade of the century, astronomers were at loggerheads over the makeup of the universe. At the instigation of George Ellery Hale, the U.S. National Academy of Sciences arranged in 1920 for Shapley and Curtis to air their views at its annual spring meeting. By chance, Shapley was visiting Berkeley, so Hale's invitation reached him there.

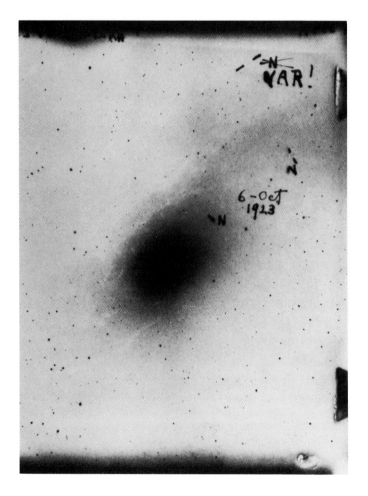

97. Hubble discovers a Cepheid outside the Milky Way. Edwin Hubble's observation logbook records the first clear evidence that the spiral nebulae were indeed island universes lying far outside our Milky Way galaxy. In his notation for photographic plate 335, taken on October 5/6, 1923, Hubble says he found a Cepheid variable in the outer regions of M31, the great spiral nebula in the constellation Andromeda. The faintness of the Cepheid meant that it was at a great distance.

98. Hubble's discovery plate. On the glass side of this negative photographic plate Hubble marked his discovery of a Cepheid variable in M31. The star was originally labeled "N" for nova, but when it reappeared on subsequent plates, Hubble realized that it must be a repeating variable star and so changed it to "VAR!"

99. Outer region of M31. This photograph, taken with the 100-inch reflector at Mount Wilson, shows objects in the galaxy's spiral arms similar to their counterparts in the Milky Way. The numerous individual points are rare highly luminous stars; those as faint as the sun cannot be seen individually but contribute to the background glow. Hubble used this photograph as the frontispiece to his popular 1936 book *The Realm of the Nebulae.*

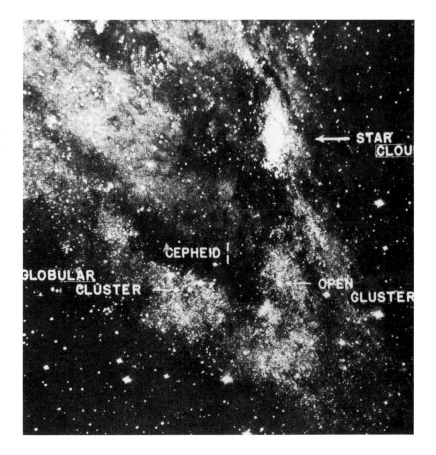

100. Cepheid variables in M31. The photographs below, showing one region of the Andromeda nebula, were taken on August 24, 1925, and November 26, 1924, respectively. They are centered on the open star cluster shown in the top picture on this page but are rotated 90° clockwise. Several variable stars appear in the pair of plates; for some, a difference in brightness is clearly distinguishable.

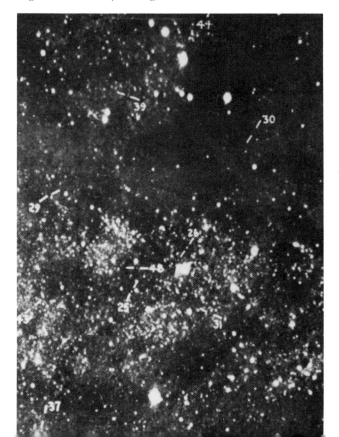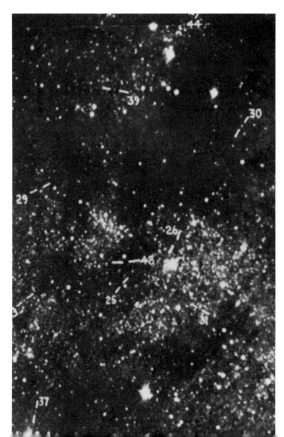

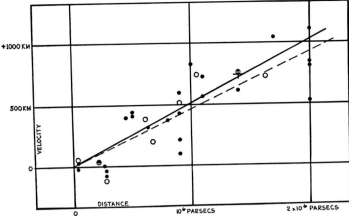

102. Hubble's velocity-distance relation. In the late 1920s Hubble made his greatest discovery. Building on his determinations of the distance to several galaxies, and on velocity measurements, he showed that the farther galaxies are, the faster they are receding. Hubble's graph shows a direct relation between a nebula's speed in kilometers per second and its distance in megaparsecs. This led to the concept of the expanding universe.

103. The depths of space. According to Hubble, this photograph, made in March 1934 with the most sensitive photographic emulsion that had ever been used with the 100-inch reflector at Mount Wilson, showed the faintest objects recordable at that time. The arrow marks an example of the faintest objects identifiable as nebulae; Hubble estimated the magnitude at 21.5 and the distance at 500 million light-years. He commented, "The plate records fully as many recognizable nebulae as stars."

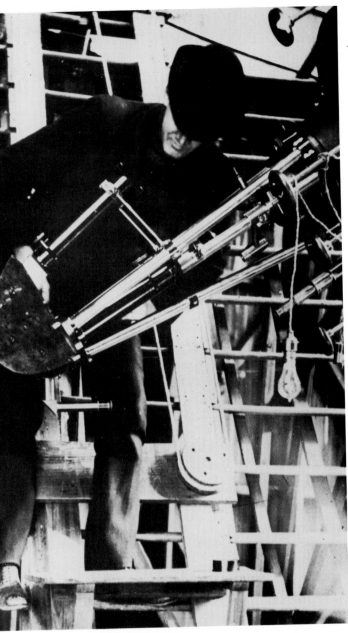

101. Slipher: the speeding nebulae. Percival Lowell, who founded an observatory for planetary studies in Flagstaff, Arizona, supposed spiral nebulae to be solar systems aborning. Thus he asked staff astronomer Vesto M. Slipher to examine them. This photograph of Slipher at the Lowell spectrograph was taken around 1912, the year he obtained the first spectrographic measurements of the astonishingly high velocities of spiral nebulae.

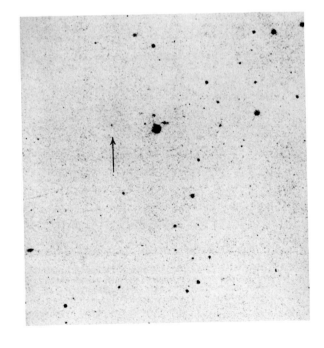

THE VELOCITY-DISTANCE RELATION
FOR EXTRA-GALACTIC NEBULAE

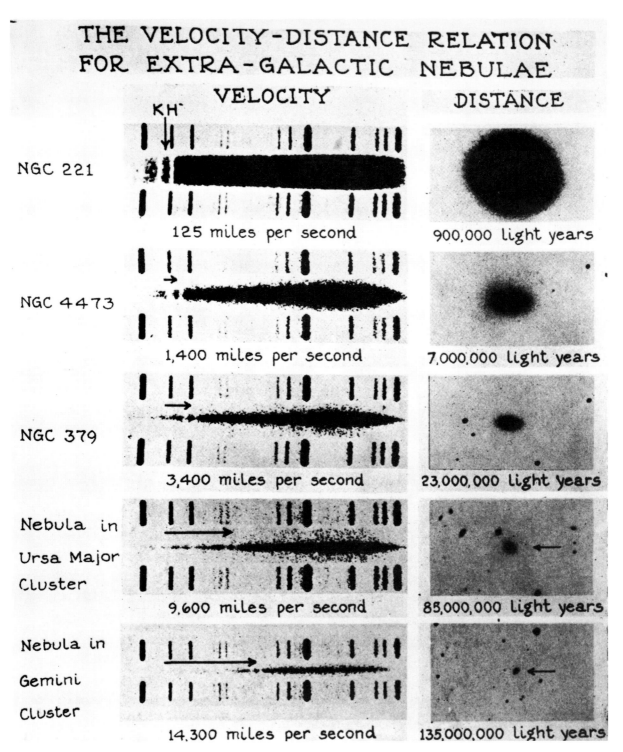

VELOCITY   **DISTANCE**

KH

NGC 221   125 miles per second   900,000 light years

NGC 4473   1,400 miles per second   7,000,000 light years

NGC 379   3,400 miles per second   23,000,000 light years

Nebula in Ursa Major Cluster   9,600 miles per second   85,000,000 light years

Nebula in Gemini Cluster   14,300 miles per second   135,000,000 light years

104. The redshift and the expanding universe. Hubble used these five spectra in the 1930s to illustrate his law that Doppler redshifts in nebular spectra increase with the apparent faintness of the nebulae. The velocities here are deduced from the Doppler shifts of the so-called H and K absorption lines that arise from ionized calcium. (The arrow above each spectrum points to the H and K lines and shows how much they are displaced toward the red.) Estimates of the distance to the Virgo cluster (represented here by NGC 4473) subsequently increased nearly tenfold, giving a much smaller slope to the velocity-distance relation, and a much older universe as reckoned from its expansion rate.

105. Barnard's dark nebulosities. Edward Barnard was one of the best early celestial photographers. Eventually obtaining a special 10-inch (25-centimeter) lens, he began recording the Milky Way, and in the process catalogued nearly two hundred dark obscuring nebulosities. This region in the constellation Ophiuchus, full of dense clouds of interstellar dust, was photographed from Mount Wilson in 1905.

7

Stars and Interstellar Matter

By the close of the nineteenth century, a few physically oriented astronomers had begun to speak of "the new astronomy" or "astrophysics." A newly formed association took the name "American Astronomical and Astrophysical Society," and entrepreneur-astronomer George Ellery Hale teamed with his American colleague James Keeler to found the *Astrophysical Journal*. This new astronomy concentrated on the physical nature of stars, especially the sun, whose surface could be examined, and it depended largely on spectrographic techniques. At the same time, large-scale photographic surveys made it possible to consider stellar motions and distributions on a statistical basis.

The spectrographic key to the physical understanding of both stars and nonstellar matter lay in the 1859 discovery, by the German physicist Gustav Kirchhoff, that the dark lines in stellar spectra could be matched against bright laboratory lines for specific elements. This was the birth of astrochemistry. Not only was it possible to determine the abundances of chemical elements in the distant stars, but eventually the same spectral lines were found to contain important clues regarding the temperature and pressure in the spectrum-forming regions of the stellar atmospheres. While glowing gas produced bright lines, cooler gas hiding a source of light and having a continuous spectrum (such as the sun, covered by its atmosphere) yielded dark lines. By the turn of the century astrophysicists had distinguished between

nebulae that were actually gaseous, those that shone by starlight reflected from interstellar dust, and those that were apparently vast, unresolved aggregates of actual stars.

Jacobus Cornelius Kapteyn, a Dutchman whose Astronomical Laboratory had no telescopes but collected photographic survey plates from other observatories for study, made one of the first major discoveries that exploited the great flood of new data. He found that the motion of stars in the Milky Way was not random but exhibited preferred directions, termed "star streaming." This motion, reported in 1904 at an international symposium at the Louisiana Purchase Exposition in St. Louis, remained mysterious for over two decades until Dutch and Swedish astronomers showed that it was a natural consequence of the rotation, like a pinwheel, of our Milky Way galaxy.

A second great discovery that rested on the flow of new results came from the Danish astronomer Ejnar Hertzsprung and the American astrophysicist Henry Norris Russell. They independently showed, around 1910, that stars did not come in a random assortment of colors and intrinsic brightnesses, but fell into specific patterns that could be delineated on a plot that has come to be called the Hertzsprung-Russell diagram. Their work supported the existence of dwarf and giant stars, a theory that was observationally confirmed with the 100-inch reflector on Mount Wilson.

At Harvard Observatory, Edward C. Pickering, who had been recruited as director from the physics department at the Massachusetts Institute of Technology, undertook the most ambitious programs for determining stellar magnitudes and spectral types, using a classification devised under his supervision by the observatory's women astronomers. The resulting great block of spectral information paid its dividend in a third great discovery based on a combination of physical insight and a vast array of data—namely, the finding by Cecilia Payne in 1925 that the apparent differences in stellar spectra arose primarily from different temperatures rather than different chemical abundances. Her calculations depended on physical formulas such as Planck's law and the notions of atomic excitation and ionization, and they yielded the then unbelievable and unacceptable result that hydrogen, the lightest element, was the predominant component of the outer layers of stars. Not until several years later did Russell persuade the astronomical community that stellar atmospheres were really composed primarily of hydrogen. Further calculations, in the 1930s, convinced astronomers that the universe as a whole must be predominately hydrogen and secondarily helium.

Meanwhile, astronomers had become increasingly aware of the problem of the source of stars' energy. How were the stars able to continue shining throughout lengthy geologic eras? The discovery of radioactivity seemed to point to unexplored resources in the nuclei of atoms, and the idea of converting mass to energy according to Einstein's $E = mc^2$ equation gradually became acceptable even in the absence of details as to how this might take place. The theoretical work by the Englishman Arthur S. Eddington, culminating in his landmark *Internal Constitution of the Stars* (1926), pointed to an intense energy source concentrated in the cores of stars. In the early 1930s, when astrophysicists realized that hydrogen would be the most abundant fuel, they suggested that the conversion of hydrogen to helium, probably via some cyclic process, could provide the required source for the sun's enormous light output. By 1938 the German physicists C. F. von Weizsäcker and Hans Bethe (by then an émigré in the United States) independently proposed that a cycle involving carbon, nitrogen, and oxygen catalyzed the conversion. (Today the "CNO cycle" is assumed to account for the normal energy production of stars more massive and hence more luminous than the sun.)

Until the emergence of astrophysics as the most exciting area of astronomical research, planets and their motions had been in the ascendancy. When Russell and his Princeton colleagues published an astronomy textbook in 1926–1927, it was the first time that a basic text devoted almost as much space to stars and nebulae as to the solar system. Yet along with the new popularity of stars and stellar systems, some interesting observations of solar system objects took place, paramount among them being the discovery of a new planet, Pluto, in 1930. Vying with the new planet in popular interest during the opening decades of the century was, of course, the apparition of Halley's Comet in 1910, although many public observers probably mixed it up in memory's eye with the even more spectacular "Great January Comet" that had appeared a few months earlier.

During these same decades, studies of another form of nonstellar matter gradually came into its own: the vast clouds of interstellar gas and dust, occasionally luminous but more generally dark. The most stunning discovery was the realization in 1930, by the Swiss-born astronomer Robert Trumpler, who was studying star clusters at Lick Observatory, that the space within the Milky Way was permeated by light-absorbing dust. Prior studies that had assumed completely transparent space were subject to revision, and the dusty universe has provided a challenge to astronomers ever since.

Owen Gingerich

106. Gauging a supergiant. The theory of dwarf and giant stars, which made its debut in the early teens of the century, predicted that a few highly luminous stars would be immense. To verify such a claim, a 20-foot, or 6-meter, interferometer designed by A. A. Michelson was mounted on the 100-inch telescope, thus extending its resolving power twenty-four fold. In 1920, in the first direct measurement of a stellar diameter, the Mount Wilson astronomers found that Betelgeuse was larger than the earth's annual orbit around the sun.

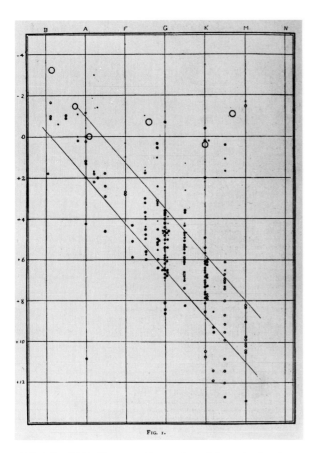

Fig. 1.

107. The H-R diagram. Reproduced here is the American astrophysicist Henry Norris Russell's first published diagram (from 1914) illustrating the relationship between stars' spectral types and luminosities. Along the vertical axis are absolute magnitudes; the letters along the horizontal axis are spectral types, from blue to red. Such graphs came to be known as Hertzsprung-Russell, or H-R, diagrams, and are used in theories of stellar evolution.

108. Hertzsprung and Russell. Soon after 1910, the Danish astronomer Ejnar Hertzsprung and Russell independently concluded that, except for rare intrinsically bright giant stars, the redder the star, the lower its absolute brightness. In this portion of a photograph from the 1913 meeting of the International Solar Union, held in Bonn, Hertzsprung is at top left, Russell at middle right.

109. The Harvard star classification team. One of the feats of the women hired as "computers" by the Harvard College Observatory was to identify variable stars in photographs of millions of ordinary stars. Annie Jump Cannon (*fifth from left*) discovered over three hundred variables. She refined the system of spectral types and classified the spectra of hundreds of thousands of stars. This group picture dates from 1925, when Cannon was curator of Harvard's astronomical photographs.

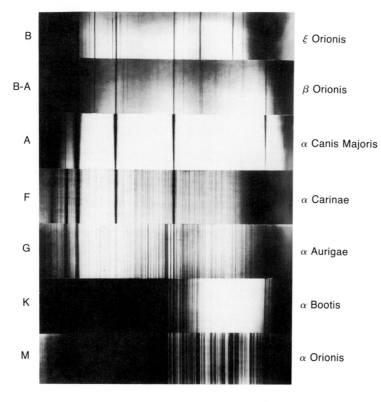

B	ξ Orionis
B-A	β Orionis
A	α Canis Majoris
F	α Carinae
G	α Aurigae
K	α Bootis
M	α Orionis

110. Spectral types. Cannon, following on the work of Williamina Fleming, showed that the spectra of most stars fell into a few types, which could be arranged in a continuous series. Reproduced here is the frontispiece to her Harvard classification of bright southern stars; it shows spectra of representative stars for the sequence of types. This scheme reflects the surface temperatures of the stars, those at the top being hottest.

111. Eddington's mass-luminosity curve. In 1924, Arthur Eddington produced an empirical diagram showing the relation between the luminosities of stars and their masses. For most stars, the greater the mass, the greater the absolute magnitude. Eddington, perhaps the most brilliant astronomical theoretician of the century, showed that this relation followed if stars behaved as gaseous spheres with a central energy source whose output depended on the central temperature. Eddington was among the first to realize that hydrogen fusion reactions provided the heat.

112. Pleiades. Seen here in a photograph made with the 36-inch Crossley reflector at Lick Observatory are the Pleiades, a nearby open star cluster in the constellation Taurus. In 1930, using Crossley plates of open clusters, the Swiss-American astronomer Robert Trumpler made the unexpected discovery that interstellar space in the Milky Way is not transparent but contains enough fine dust to absorb part of the light from distant clusters.

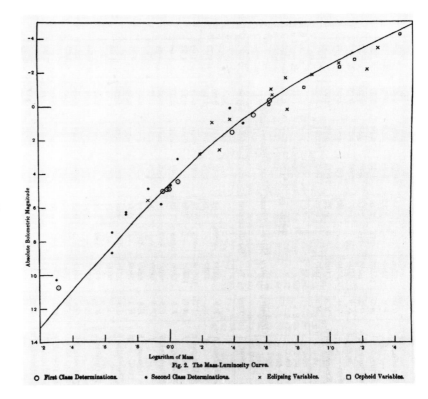

Fig. 2. The Mass-Luminosity Curve.

O First Class Determinations.   • Second Class Determinations.   x Eclipsing Variables.   □ Cepheid Variables.

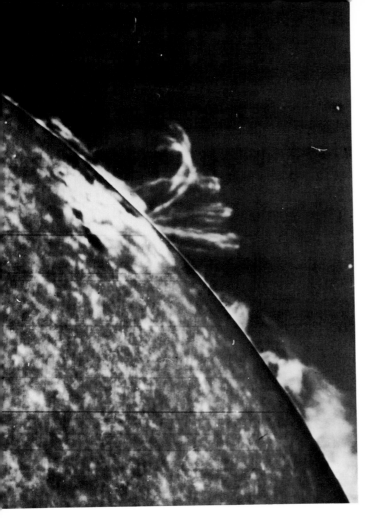

113. The sun in the light of glowing calcium. At the turn of the century, a new era in solar research was opened when George Ellery Hale in America and Henri Deslandres in France developed the spectroheliograph. Using the light of a single line in the solar spectrum, this instrument produced detailed images that were not possible with the integrated white light of the sun. While using light from calcium, Hale found luminous clouds on the sun's disk, which he called flocculi; this 1907 Yerkes Observatory picture is actually a montage, combining a spectroheliogram of erupting prominences with a spectroheliogram of the bright underlying flocculi.

114. Giant sunspot group (*lower left picture*). Hale's California observatory on Mount Wilson included two tower telescopes to give large images of the sun. The 60-foot tower telescope recorded this enormous sunspot group, many times larger than the earth, in 1917. The dark dot shows the size of the earth.

115. A simple sunspot in detail. Improvements in astronomical instruments have brought remarkably refined solar images, displaying, for example, the boiling granular texture of the sun's photosphere. This 1970s photograph from Sacramento Peak Observatory in New Mexico shows clearly the spot's filamentary penumbra.

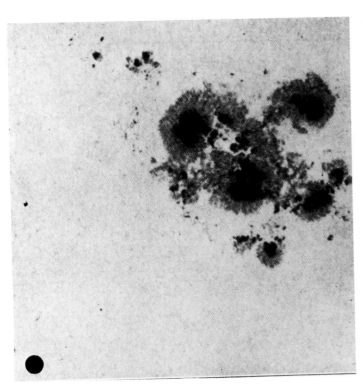

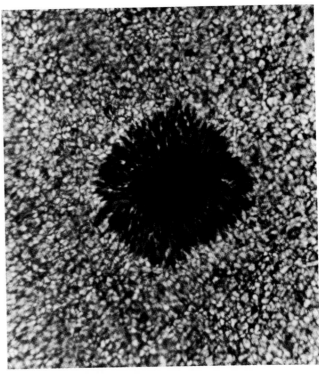

116. An early view of an astrophysical laboratory. Hale provided his observatory on Mount Wilson with a well-equipped physical laboratory. The large electromagnet and spectrograph were used to study the Zeeman effect—the splitting of a spectral line when the light source is in a magnetic field. Hale discovered the effect in some spectrograms of sunspots, and experiments were carried out in the Mount Wilson lab to aid interpretation of the astronomical observations.

117. Solar prominence. This mass of luminous gas, seen in the light of calcium, rose 225,000 kilometers (140,000 miles) above the sun's disk on July 9, 1917. The white circle indicates the size of the earth. The photograph was made with the spectroheliograph in the 60-foot tower telescope on Mount Wilson by Ferdinand Ellerman.

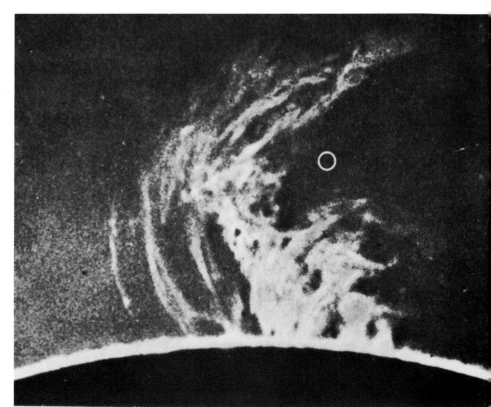

118. Investigating the "Tunguska meteorite." At 7 A.M. on June 30, 1908, a blindingly bright fireball crossed the sky and exploded over Siberia, with a sound heard for more than 1,000 kilometers. A seismic wave was recorded as far away as Western Europe, and the shock wave traveling through the air was registered on barograms in St. Petersburg and even in Great Britain. The photograph shows the first scientist to investigate the incident, the Soviet Leonid A. Kulik. He conducted expeditions to the site beginning in 1927.

119. Aftermath of the Tunguska explosion. This picture was taken by Kulik in May 1929, nearly twenty-one years after the explosion. He found that trees had been flattened up to 30 kilometers (nearly 20 miles) from the epicenter of the event. The most plausible explanation is that the observed fireball, stupendous explosion, and shock wave were caused by a small comet, or a fragment of a comet, that entered the earth's atmosphere.

120. Halley's Comet. In 1910, Halley's Comet made a memorable swing around the sun, immersing the earth in its tenuous tail. Some people viewed the approaching comet with dread, fearing deadly gases. Others in the humorous French postcard above saw a chance to strike it rich: some of those fleeing the earth are using the air belts advertised for 60 francs apiece in the upper left. At right is the comet as it really was in May 1910. The comet's return in 1986 provided scientists with a space-borne close-up look at its nucleus but left the public disappointed.

121. Discovery of a planet. The ninth planet from the sun, Pluto was found by Clyde W. Tombaugh of the Lowell Observatory in 1930, after a painstaking search. The upper plate was taken on January 23, 1930; the lower one six days later. While examining the plates, Tombaugh saw that the tiny dot, now marked by an arrow, was a moving object, which turned out to be the long-sought outer planet.

Part Three

THE EARTH
AND ITS ENVIRONMENT

Kaibab Limestone
Toroweap Limestone

Coconino Sandstone

Hermit Shale

Supai Sandstone

Redwall Limestone

Muav Limestone

Bright Angel Shale

Tapeats Sandstone

Grand Canyon Series

Zoroaster Granite

Brahma and Vishnu Schist

122. A gash through time. The twentieth century saw absolute ages established for the geologic column of strata, here exemplified by Arizona's mile-deep Grand Canyon, whose layers of rock stretch 2 billion years back, from the 250-million-year-old Kaibab Limestone on the canyon rim down into the Precambrian era of the inner gorge.

8

*T*he *A*ge
of the *E*arth

Since the seventeenth century the age of the earth has been one of the most controversial numbers in the history of science. In 1650, James Ussher, the archbishop of Armagh, Ireland, asserted that the world was created in 4004 B.C. Although other scholars disagreed on the exact date of the creation, most European scientists accepted the biblical chronology that assigned the earth, and in fact the entire universe, an age of only about 6,000 years. When the French naturalist G.-L. L. de Buffon, writing in the middle of the eighteenth century, estimated that it took 75,000 years for the earth to cool from a hot fluid initial state, this figure was considered heretical, and Buffon was ordered by the theological faculty of the Sorbonne to retract it. Yet Buffon and other scientists were already beginning to suspect that the age of the earth should be measured in millions or even hundreds of millions of years.

Nineteenth-century geologists, especially the followers of the "uniformitarian" doctrine of Charles Lyell, saw no reason to adopt any particular value for the age of the earth. Lyell, in his *Principles of Geology*, argued that the only scientific way to reconstruct the earth's history was to invoke just those causes or forces that could now be seen in operation—erosion, uplift, earthquakes, and volcanic eruptions. Lyell also assumed that these familiar forces had not been significantly more intense in the past. In particular, he insisted that the earth's surface temperature had been more or less the same during the entire period of

geologic history. The formation of the earth took place before geologic history began and could be pushed indefinitely far back into the past.

Charles Darwin, following Lyell's approach, suggested in his *Origin of Species* that periods of the order of 300 million years, estimated from the rate of geologic processes, might be available for the slow process of evolution by natural selection. But Lord Kelvin, the most influential British physicist of the day, said that the earth could not have taken more than 100 million years to cool down; he subsequently reduced his estimate of the earth's age to only 24 million years. This seemed to be a formidable objection to Darwin's theory.

After the isolation of radium by Marie and Pierre Curie, it was recognized that Kelvin's estimate was much too low. He had implicitly assumed that no heat is generated inside the earth to replace that which flows outward to the surface. The Curies' work showed that there is probably enough radium in the earth's crust to generate more heat than the amount lost by radiation into space. Thus, it seemed, scientists might have to abandon the assumption, accepted for more than a century, that the earth had been slowly cooling down since its formation; perhaps it was actually warming up.

Soon after Ernest Rutherford proposed that radium and other radioactive substances are transmuted into different elements by alpha decay (see Chapter 1), he pointed out that one could use this process to measure the ages of rocks. For example, if the helium trapped in a rock had been produced by the decay of radium or uranium at a known rate, one could determine how long the rock had been solid. Using this "radiometric" approach, he announced in 1904 that a sample of fergusonite rock was 40 million years old; this figure was soon revised to 500 million years. The British physicist J. W. Strutt estimated an age of 2.4 billion years for a specimen of thorianite. Adopting a different approach, the American chemist Bertram Boltwood used the amount of lead (assumed to be produced by the decay of uranium) to obtain a figure of 2.2 billion years for another sample of thorianite.

These early estimates were unreliable, both because some of the helium could have escaped from the rock and because some of the helium and lead might have been produced by the decay of other elements, such as thorium. After the discovery of isotopes, detailed sequences of radioactive decay could be established, and more accurate estimates made. By 1920 the Scottish geologist Arthur Holmes had established that the most ancient minerals were at least 1.6 billion years old, and he estimated that the earth's crust was probably about 2 billion years old. Holmes and others also showed that the methods of radiometric dating could be used to provide a consistent time scale for the science of historical geology.

The best method for determining the age of the oldest rocks is based on measurement of isotopes of uranium and lead, using the mass spectroscope. Lead 206 and 207 are the end products of decay series starting with uranium 238 and 239, respectively, while another "radiogenic" isotope, lead 208, comes from the decay of thorium. A fourth lead isotope is not produced by any such process and is therefore called nonradiogenic or primeval; it was presumably present at the time of the earth's formation. The American physicist Alfred Nier found that the lead isotopes occurred in different proportions in different rocks; he suggested that these variations could be ascribed to the different rates of the decays that produced the isotopes and could be used to infer the time that had elapsed since the rocks' formation.

In 1946, Holmes and the German physicist Friedrich Houtermans independently pointed out that Nier's measurements of lead isotope abundances could give not only the ages of particular rocks but also the age of the earth itself. Since the rate of radioactive decay is affected hardly at all by physical and chemical processes, it allows one to probe the period before the oldest rocks were formed, when the earth's crust may have been fluid. Holmes and Houtermans used their method to extend the earth's age to about 3 billion years.

In 1953, a group of scientists at the University of Chicago and Caltech reported a new determination of the abundances of lead isotopes in meteorites. This result suggested a baseline figure for the proportion of primeval (nonradiogenic) lead in terrestrial rocks. Clair Patterson, a member of the group, announced that the earth is at least 4.5 billion years old. During the next three years this value was confirmed by other analyses of lead isotope data and by completely independent estimates based on two other radiometric dating techniques. One of those techniques involved the decay of potassium to argon; the other employed the decay of rubidium to strontium. In 1956, Patterson was able to establish the age of the earth (and of some meteorites) at 4.55 ± 0.07 billion years.

By the 1970s almost all scientists accepted Patterson's value for the age of the earth. Analysis of lunar rocks showed the moon to be also about 4.6 billion years old, and this came to be taken as the approximate age of the entire solar system.

Radiometric dating has played a major role in the development of geology in the twentieth century. It permitted the establishment of an absolute time scale for geologic periods and provided dates for many important events in the earth's recent and distant past. For example, a major turning point in the revolution in the earth sciences during the 1960s was the discovery that the patterns of direct and reversed magnetization on the ocean bottoms could be correlated with accurately dated worldwide geomagnetic reversals. It is remarkable that even though this revolution changed much of what scientists thought they knew about earth history, the most fundamental parameter, the age of the earth itself, remained fixed: its accepted value remained within the error limits given by Patterson in the mid-1950s.

Stephen G. Brush

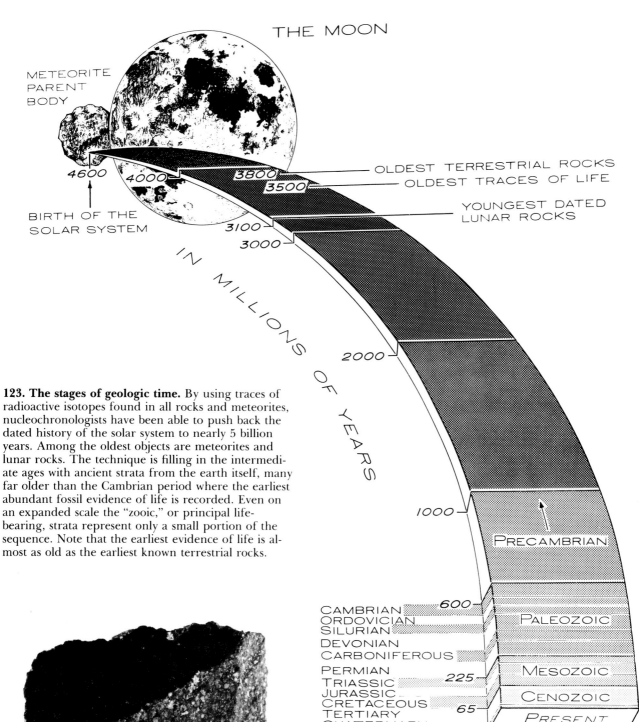

THE MOON

METEORITE
PARENT
BODY

4600 *4000* *3800* — OLDEST TERRESTRIAL ROCKS
3500 — OLDEST TRACES OF LIFE

YOUNGEST DATED
LUNAR ROCKS

BIRTH OF THE
SOLAR SYSTEM

3100
3000

IN MILLIONS OF YEARS

2000

1000

PRECAMBRIAN

123. The stages of geologic time. By using traces of radioactive isotopes found in all rocks and meteorites, nucleochronologists have been able to push back the dated history of the solar system to nearly 5 billion years. Among the oldest objects are meteorites and lunar rocks. The technique is filling in the intermediate ages with ancient strata from the earth itself, many far older than the Cambrian period where the earliest abundant fossil evidence of life is recorded. Even on an expanded scale the "zooic," or principal life-bearing, strata represent only a small portion of the sequence. Note that the earliest evidence of life is almost as old as the earliest known terrestrial rocks.

CAMBRIAN *600*
ORDOVICIAN
SILURIAN
DEVONIAN
CARBONIFEROUS
PERMIAN *225*
TRIASSIC
JURASSIC
CRETACEOUS *65*
TERTIARY
QUATERNARY

PALEOZOIC

MESOZOIC

CENOZOIC

PRESENT

124. Oldest dated macroscopic object. This fragment of the Allende meteorite, a ton of which fell in northern Mexico in 1969, carries small white nodules dated at 4.6 billion years. The nodules are pieces of the early solar system, miniature building blocks for planets, asteroids, and comets. As the meteorite hurtled through the earth's atmosphere, it fragmented, and the surfaces melted from air friction, producing the black crust visible on the top of this sample.

125. Rock collecting on the moon. Beginning in 1969, several missions in the U.S. Apollo program took astronauts to the surface of the moon. There they collected samples of rock for study, opening up an exciting new field of investigation for geologists. In this picture from the February 1971 Apollo 14 mission, a hammer and a small collection bag lie on top of a lunar boulder. Samples of lunar material were also obtained, on a much smaller scale, by unmanned Soviet missions.

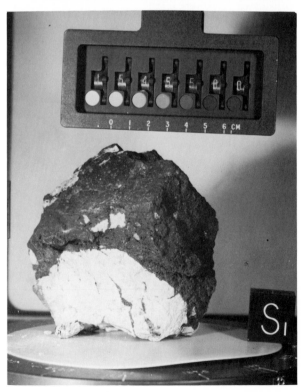

126. Ancient lunar rocks. These two samples were collected in 1971 by the Apollo 15 astronauts. At right, from the rim of a crater near the Apennine Mountains, is one of the oldest moon rocks, dated at 4.5 billion years. This fine-grained specimen was apparently formed in the heat and pressure generated by a meteorite impact. The sample above, found in the gorge known as Hadley Rille, is a vesicular olivine basalt. The holes, or vesicles, resulted when gas bubbles were frozen in the lava as it solidified. Dated at about 3.4 billion years, it is one of the younger rocks on the lunar surface.

127. Ancient rocks in Greenland.
In 1972 the mountains in the background of this photograph were dated as mainly 3.7 billion years old, making them appreciably older than any other rocks then known on earth and hence the oldest part of the earth's crust whose age could be reliably established. The rock layers are the so-called Amitsoq gneisses. In 1983, zircons dating from over 4 billion years ago were found in Western Australia.

128. Oldest remains of life. As of the early 1980s the oldest traces of life had been found here: the Pilbara Block of the Warrawoona Group in the "North Pole" region of Western Australia. The age: about 3.5 billion years. These are fossil layers of blue-green stromatolites, procaryotic algal mats that covered shallow tropical waters in the earth's earliest seas.

129. Ancient sedimentary rock. This sample of the so-called banded-iron formation from Greenland, dated at 3.76 billion years in age, is among the oldest known sedimentary rocks on earth. The dark bands are rusted iron, which must have periodically locked up the oxygen in the atmosphere—a situation that prevailed for hundreds of millions of years.

130. Ediacaran fauna. Here are examples of the surprisingly complex fossils that geologists began finding in the 650-million-year-old Ediacara sandstones in South Australia in the 1940s. These very early soft-bodied marine organisms—some of which may represent unsuccessful phyla, unrelated to present forms—have given their name to a Precambrian geological period.

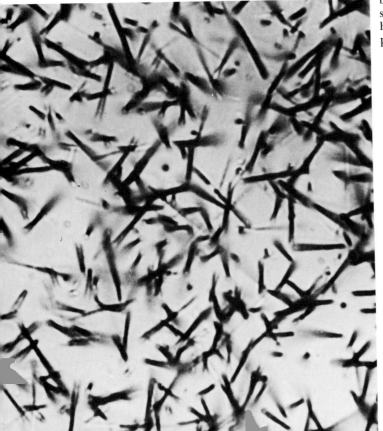

131. Dating rocks by fission tracks. This crystal of zircon, under magnification, reveals a multitude of tiny tracks caused by the spontaneous fission of uranium 238 atoms, which are present as an impurity in many rocks. By measuring the number of fission tracks left in a sample, and comparing it with the sample's uranium content, the age of the rock can be determined. This technique, developed after 1959, can use a very small sample and can cover a broad span of time.

132. Radiocarbon dating. After World War II the University of Chicago chemist Willard Libby discovered that the age of organic objects, including archaeological artifacts such as wood or cloth, could be measured by analyzing the proportion of radioactive carbon 14 in them. The method can establish ages up to about 50,000 years. The photograph, taken in 1957, shows a technician preparing a pump for purifying the carbon dioxide gas from the specimen she is analyzing.

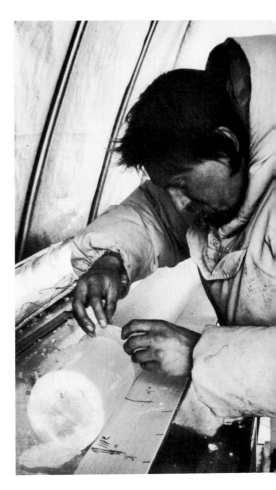

133. Ice core. By drilling deep into ice sheets, scientists can retrieve ice made tens of thousands of years ago and can date it by counting the annual layers. The ice layers visible in a core extracted from a borehole reveal much about the history of climate—a thicker layer, for example, might mean a heavier snowfall, or a layer of ash could reflect a volcanic eruption. Here a Greenland ice core is illuminated by the light table as a scientist studies its layers.

134. Egyptian funerary ship. This cedar boat from the tomb of Pharaoh Sesotris III, shown in a turn-of-the-century exhibition in Chicago's Field Museum, provided one of the samples used to establish the radiocarbon dating method. The boat was known to date from about 1800 B.C. When the cedar tree was cut down, its intake of new carbon, including carbon 14, ceased. Since the carbon 14 decayed with a known half-life of 5,730 years, the age of the cedar could be estimated by measuring its present proportion of radioactive carbon. The technique gave as of 1950 A.D. a result of 3,621 years plus or minus 180.

135. The *Trieste*. After reaching unprecedented heights in balloon ascents in the early 1930s, the Swiss physicist Auguste Piccard turned to the ocean depths. His bathyscaphes, unlike the bathysphere of the 1930s, were navigable, their buoyancy could be controlled, and they could reach much greater depths. In 1953, Piccard and his son, Jacques, trebled the existing depth record. Five years later the U.S. Navy bought and reequipped the *Trieste*. In 1960, Jacques Piccard and U.S. Navy Lieutenant Don Walsh took it to the deepest part of the Pacific, to a record 10,900 meters (36,000 feet) in the Marianas Trench. The later, smaller research submersibles were much more maneuverable but could not operate at the great depths plumbed by the *Trieste*.

9

Oceanography

By 1895, when the last volume of the famous *Challenger* expedition (1872–1876) was published, isolated bits of information about the world's oceans had been fashioned into a picture that would accommodate further oceanographic discoveries for more than fifty years—for the years, that is, until oceanographers and geophysicists adopted the hypothesis of seafloor spreading and radically revised science's understanding of the earth's history.

As the twentieth century began, however, oceanography was curiously directionless. Although a vast realm had been opened up for exploration, governments on both sides of the Atlantic were loathe to finance a science with so few practical advantages. In Europe the need for food fish soon spurred organized marine studies of coastal regions, but in the United States the science remained the province of imaginative amateurs and of academics who operated summertime laboratories. A fascinating prospect for both groups was the direct observation of deep-sea life, and as early as 1934 the American zoologist William Beebe and engineer Otis Barton descended 3,000 feet in a bathysphere. Beebe wrote popular books about the strange and beautiful creatures he saw. After World War II the Swiss scientist Auguste Piccard was prompted by a similar urge to sample the depths directly. He developed the more sophisticated bathyscaphe *Trieste,* which was used in 1960 to explore one of the ocean's deepest trenches. The 1960s began a new period of worldwide explora-

tion, as several notable deep-diving research submersibles were launched, including the American *Alvin.*

Meanwhile, progress had been made in mapping the contours of ocean basins, an endeavor that could not proceed efficiently until isolated deep-sea soundings made with weighted rope or wire were replaced by continuous, acoustic measurements. The development of echo sounders was spurred first by the need to detect submerged icebergs (after the British liner *Titanic* collided with an iceberg in 1912 with a loss of 1,500 lives) and then by the need to detect enemy submarines during the two world wars. Early sonic detectors sent beams of sound horizontally, and listeners timed the returning echoes in order to locate submerged objects such as submarines. When the beams of sound emitted by these devices were aimed toward the seafloor, the detectors became, in effect, echo sounders. With them, continuous recordings of the ocean bottom were made.

Details of the ocean basins accumulated rapidly, but by themselves they were not enough to challenge the traditional belief that oceans and continents were permanent, fixed features of the earth's crust. It was the exploration of the structure of the seafloor itself—in particular, the determination of the thickness and age of its sediments—that suggested that oceans were evolving features and that the seas we look upon today are relatively recent.

If oceans had since the beginning of time received the debris washed from continents, then the marine sediments overlying the earth's basaltic crust should be at least several miles thick. The deeper one sampled within these layers, the older the material should be. Sediments on the crust should be several billion years old—the age of the most ancient continental rocks. But how could the thickness and age of sediments lying miles beneath the surface be measured?

After World War II, oceanographic vessels began using surplus explosives to examine the structure of the seafloor. With a procedure known as seismic profiling, powerful bursts of sound were sent through the water and through the ocean sediments. Hydrophones picked up a series of echoes as the sound returned first from the seafloor and then, soon after, from sediment layers beneath it and from the earth's crust. Under the determined efforts of such American geophysicists as Maurice Ewing, seismic profiling quickly became more sophisticated, and by the 1960s data suggested that there was much less sediment than expected in ocean basins. During the same years deep-sea drilling revealed that ocean sediments were relatively young. Scientists used microscopic marine fossils as indicators of the sediment's age and found the oldest oozes to be only about 150 million years old. Mapping the earth's gravitational and magnetic fields over the deep sea added still more information concerning crustal structures.

The new data supported the concept of seafloor spreading and the broader theory of plate tectonics (both described in the next chapter), and continents and ocean basins were now seen by such pioneers as the American geologists Harry Hess and Bruce Heezen and the British geophysicists Frederick Vine and D. H. Matthews as dynamic, constantly evolving features. This reformulation constituted a revolution in ideas concerning the history of the ocean basins and continents.

A second revolution in the earth sciences was triggered by the discovery of a seafloor ecosystem whose food chains are independent of the sun and whose members subsist on sulfur and other chemicals. In the late 1970s unmanned submersibles loaded with photographic and temperature-sensing equipment and towed from surface ships brought back puzzling data from the East Pacific Rise. The *Alvin* went down to look directly at what seemed to be isolated regions of extremely hot water. It brought back dramatic photographs of "smokers" spewing clouds of hot water over surrounding colonies of giant clams, tube worms, and unknown dandelion-shaped organisms. Gradually scientists realized that in some volcanic areas water circulates below the seafloor. Vented through chimneylike formations, the hot water gushes from the seafloor bringing with it heavy loads of dissolved minerals. The American microbiologist Holger Jannasch and others pulled the information together and hypothesized that the strange organisms had evolved in concert with specialized bacteria that make use of these materials. The organisms derive their energy from chemosynthesis rather than photosynthesis.

The circulation of ocean waters is a third area of research in which great strides have been made in this century. Using both theoretical modeling and direct observations from ships, buoy arrays, and satellites, scientists gained a general understanding of how ocean circulation is organized worldwide. The Swedish meteorologist Carl-Gustaf Rossby, the American oceanographer Henry Stommel, and their associates developed mathematical models of ocean circulation that explained, among other things, why strong, warm currents such as the Gulf Stream hug the western boundaries of ocean basins.

At the same time, new techniques of direct observation carried far forward the work that Harald Sverdrup and other Scandinavian oceanographers had done on computing ocean currents from measurements of temperature, salinity, and depth. Using satellites and buoy arrays, physical oceanographers could now take closely spaced and nearly simultaneous measurements of water temperature, color, texture, and topography. With this information they could monitor not only long-term phenomena such as the oceanic response to the gradual atmospheric warming known as the greenhouse effect but also relatively short-term phenomena such as storm surges and tsunamis (tidal waves). Although much work in physical oceanography has been theoretical, practical advantages have accrued, including improved long-term weather forecasts and a clearer understanding of the consequences of using the sea as a dumping ground.

Susan Schlee

136. Beebe's bathysphere. Constructed by the American zoologist William Beebe and the engineer Otis Barton, the bathysphere was a hollow steel ball suspended by cable from a ship. Beginning in 1930, Beebe used the two-passenger sphere to observe deep-sea life. By 1934, Beebe and Barton had reached a depth of nearly 900 meters (3,000 feet).

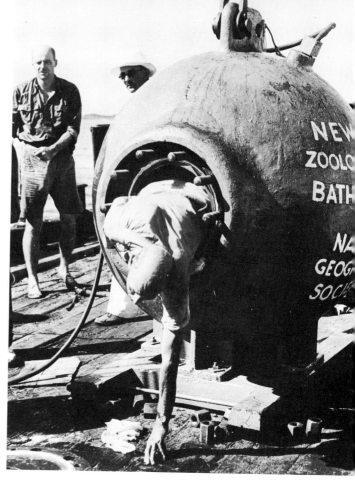

137. Alvin. The 1950s saw the first use of small self-propelled, manned research submersibles for undersea exploration. The U.S. submersible *Alvin*, built in the 1960s, became the best-known deep-diving vehicle. Here we see *Alvin* with a new titanium pressure hull it received around 1974, which allowed it to work at depths as great as 3,700 meters (12,000 feet).

138. Alvin returns to Lulu. Here, *Alvin* is returning to its catamaran support tender. The 105-foot *Lulu* has laboratory and work space for several scientists, along with berthing and mess facilities and a repair shop. *Alvin* itself can accommodate three crew members. It is equipped with navigation instruments, television and camera systems, and a mechanical manipulator capable of lifting up to 50 pounds of bottom samples. When not diving, it is normally carried on board *Lulu*.

139. Manganese nodules. Such potato-shaped concretions, formed primarily of manganese salts and manganese oxide minerals, were first dredged up by the expedition of the H.M.S. *Challenger* in the 1870s. First photographed in 1948 near Bermuda at a depth of 5,500 meters (18,000 feet), they are apparently formed by sedimentation and precipitation processes, often around seashells. Trillions of tons of the nodules exist on the floors of the world's oceans and are potentially a major mining source of manganese. U.S. and Japanese experimental mining of the nodules began in 1970.

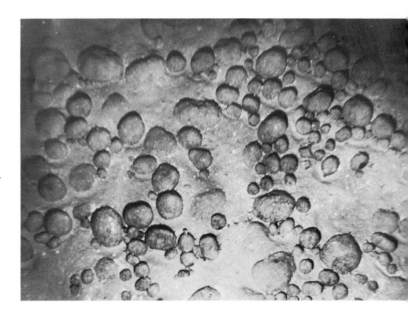

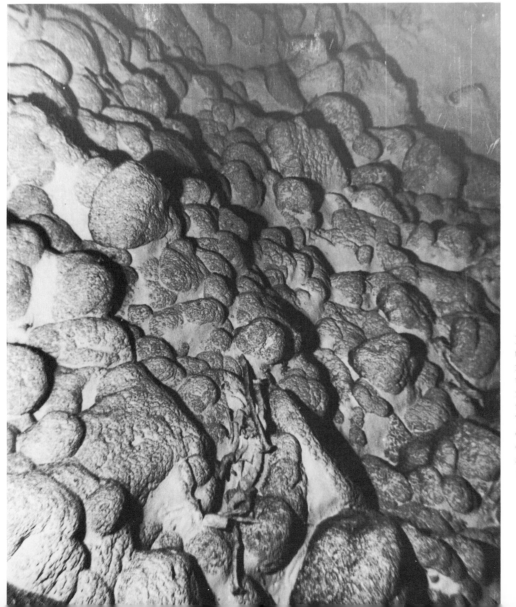

140. Pillow lava on the ocean flo
Researchers exploring the floor o
the mid-Atlantic found curious ro
formations like these. Lava wellin
up from below often assumes
rounded shapes like pillows owing
to rapid cooling in the ocean wate
The lava shown here was observe
on the western slope of the mid-
ocean ridge in the South Atlantic
a depth of 2,700 meters (8,900
feet).

141. Colony of dead giant clams. The deep-ocean floor was long thought to be a barren realm of mud and rock. Since sunlight provides the energy that powers the food chain, how could it be otherwise? But in the late 1970s explorers began finding on the deep-ocean bottom colonies of organisms unlike any previously known. These depend for their sustenance on chemosynthetic bacteria, which obtain needed energy from hydrogen sulfide gas and other substances released by hot vents. Visible in this 1978 photograph from the East Pacific Rise is the mechanical arm of the French submersible *Cyana*.

142. Tube worms. Among the strange creatures found near the deep-sea hot springs, or hydrothermal vents, are these giant tube worms, which have lengths of up to several feet. This unexpected oasis was discovered in 1977 by geologists on the *Alvin* in the area of the Pacific bottom known as the Galápagos Rift. The flash photograph shown here was taken in the dark waters about 2,600 meters below the surface early in 1979.

143. Hydrothermal vent. The temperature of the superheated water issuing from this vent on the ocean floor was measured at 300° Celsius. The chemicals carried by such smokers would be poisonous to most forms of life, but colonies of exotic fauna and chemosynthetic bacteria thrive in the vicinity of suboceanic hot springs. This 1982 photograph was taken on an expedition of the *Cyana*.

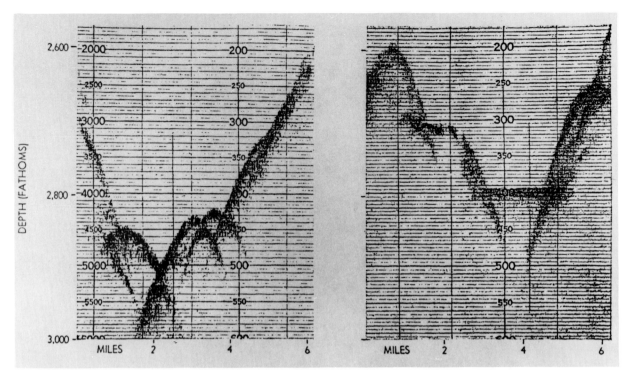

144. Echo sounding. Detailed mapping of the ocean bottom was not possible before echo sounders were developed between the two world wars. The introduction in 1954 of the precision depth recorder, accurate to 1 meter in 3,000, made possible extensive detailed mapping of the deep sea floor. Reproduced here is a cross section of part of a chain of deep, narrow trenches circling the central basin of the Pacific. The cross section at left, made near Acapulco, reveals a V-shaped bottom and minimal sediment; the depth is about 5,000 meters (16,000 feet). The right cross section, near Manzanillo, shows a flat bottom.

145. Seismic profile of the Atlantic floor. This cross-sectional view of the western flank of the Mid-Atlantic Ridge is a "continuous seismic profile" obtained in 1967 by the American vessel *Stranger*. Seismic profiling, developed in the 1960s, is a kind of super echo sounding. The ship tows a sound source and a string of underwater hydrophones for detecting seismic echoes from rock layers. The method can reveal the structure of the sediments at the ocean bottom and the shape of the basaltic basement below them. In the area profiled here, the sediments are relatively flat and uniform. The basaltic basement protrudes through the sedimentary layers at left and comes to dominate the topography closer to the ridge.

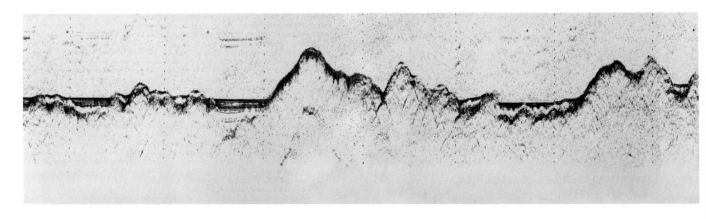

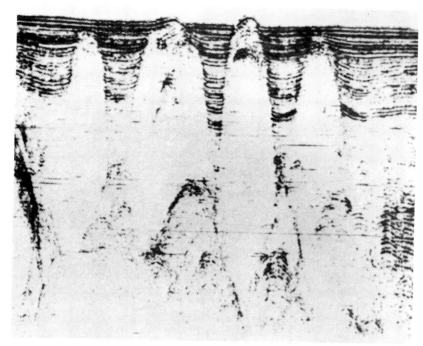

146. Seismic profile of the Mediterranean floor. These pillars, discovered in seismic profiling in 1961, appear to be salt domes a few miles in diameter and rising hundreds or thousands of feet high. Some protrude through the sediments on the sea floor and form knolls. This was the first indication that huge deposits of salt lie under the floor of the Mediterranean, evidence of extreme desiccation in earlier geological times.

147. Gulf Stream in the infrared. The development of artificial earth satellites opened up a major new source of data for oceanographic study. This view of the Gulf Stream taken from a U.S. weather satellite uses infrared light to show the Gulf Stream's warm waters flowing to the northeast.

148. Mohole fiasco. In 1957, American scientists inaugurated a project to drill a hole through the earth's crust and into the underlying mantle. The crust's lower bundary, called the Mohorovičić discontinuity, or Moho, varies in depth from more than 40 kilometers (25 miles) under mountains on continents to as little as 5 kilometers under the ocean. Phase I of Project Mohole, carried out by the drilling ship *CUSS 1* shown here, broke the world record for deep drilling. But reaching the Moho was a staggering problem for drilling technology then, and the project was killed by poor planning and political snafus.

149. *Glomar Challenger.* In 1968 this 400-foot American drilling vessel first put to sea. The night view from the bridge (*left*) shows the lower part of the 142-foot drilling derrick and, beyond, an automatic pipe racker for storing 23,000 feet, or 7,000 meters, of drill pipe. Suspended from the derrick is the top part of a drill string 15,000 feet long. The picture was made during the drilling of a borehole in the Mediterranean west of Sardinia. In the photograph at right, we see rock salt retrieved from more than 10,000 feet below sea level. Such salt could have been deposited from the extremely bitter waters of an evaporating Mediterranean, and indeed the 1970 drilling yielded strong evidence that the sea had dried up 6 million years ago, producing a giant desert.

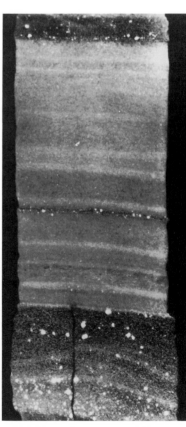

150. World's deepest thirty-story-high derr protects a giant Soviet rig to drill 15,000 meters (about miles) into the earth's crust. Dril began here, a site on the Kola Peninsula near Norway, in 1970; by 1984 the well had reached 12,000 meters. The project was intended to yield both scientific knowledge and practical information relating to the crust's deep structure and to deeply-lying mineral deposits and oil and gas fields.

151. Kola wellhead. Here is a view inside the drilling derrick that supports the drill string. Flanking the wellhead are sections of aluminum-alloy drill pipe. The derrick housing kept the wellhead above freezing, permitting year-round operations even though the drill site was north of the arctic circle.

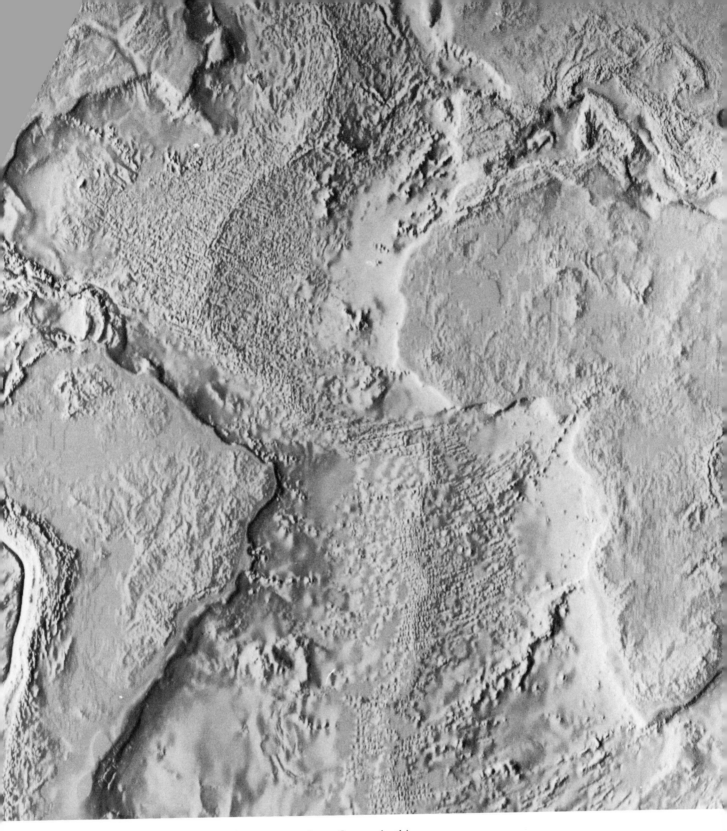

152. Atlantic Ocean bottom. The most prominent feature in this computer-generated topological map is the ridge and accompanying rift valley running down the middle of the ocean floor. According to the theory of plate tectonics, new crust is constantly formed from below at the mid-ocean ridges and spreads to each side of them.

10

Continental Drift

Ever since reasonably accurate maps of the earth were first published in the sixteenth century, the rough parallelism of the coastlines of the continental masses bordering the Atlantic Ocean suggested that those masses were once joined together. But the hypotheses proposed to explain the separation of the continents generally involved catastrophic events that seemed to be beyond the reach of scientific investigation.

The modern, noncatastrophic theory of continental drift was first developed by Alfred Wegener, a German meteorologist trained also in astronomy. He was impressed not merely by the congruence of the Atlantic coastlines but also by the similarity of fossil biological species on both sides of the ocean. Paleontologists explained this similarity by postulating prehistoric "land bridges" across the Atlantic. Wegener found this improbable and preferred to assume that the continents had once all been part of a single giant landmass, which he called Pangea. He collected a large amount of biological and geological evidence to support this hypothesis, which he presented in a series of articles in 1912 and in a book, *The Origins of Continents and Oceans*, in 1915.

According to Wegener, the continents are blocks of light granitic rock that float on denser basaltic rock, which is exposed on ocean floors. They are acted on by two forces. First, because of the departure of the earth from spherical shape, the gravitational and buoyant forces do not precisely balance but produce a small net force pushing the

continents toward the equator. Wegener credited the discovery of this force to the Hungarian physicist Roland Eötvös. Second, the tidal attraction of the sun and the moon exerts a drag on the crust, slowing its rotation relative to the inside of the earth and producing a net westward drift.

Wegener's theory was generally rejected by scientists in the 1920s and 1930s. A major objection was that the Eötvös and tidal forces are much too weak to push solid continents through the solid layer beneath them. According to the views of Lord Kelvin, generally accepted by scientists at the beginning of the twentieth century, the entire earth is solid, "as rigid as steel."

The plausibility of any theory about the motions of the earth's outer crust depends both on direct evidence for such motions and on the current state of knowledge about the properties of the interior. In particular it relies heavily on seismology, the science of earthquakes. Although earthquakes had been studied for several centuries, the development of systematic quantitative seismology began in Japan in the late nineteenth century. The Meiji Restoration of 1868 led to the reopening of Japan to Western scientists; among them were John Milne, C. G. Knott, J. A. Ewing, and Thomas Gray, who came from Britain to teach in Japanese colleges. Following an earthquake at Yokohama in 1880, Milne and several of his British and Japanese colleagues founded the Seismological Society of Japan, the first organization of its kind in the world. Milne developed a compact seismograph that could be installed at observing stations widely distributed over the surface of the earth, thus making it possible to collect the reliable global data needed to analyze the propagation of earthquake impulses through the earth's interior.

At the same time Ernst von Rebeur-Paschwitz was developing a very sensitive seismograph in Germany; in 1889 he was able to detect at Potsdam and Wilhelmshaven the impulses from an earthquake in Tokyo. This was the first observation of seismic waves that had passed directly through the earth. A few years later Richard Oldham, an Irish geologist who directed the Geological Survey of India, began analyzing records from several stations of earthquakes in India, Argentina, and Guatemala. Emil Wiechert, a physicist at the University of Göttingen, designed a new seismograph, established a network of observatories in the German colonies, and developed theoretical methods for inferring the earth's structure from seismic records.

By 1940 seismologists had established the overall picture of the earth's structure. The outer crust extends as deep as 40 kilometers (25 miles) under the continents but only about 5 kilometers under the ocean floors; it is separated from the solid upper mantle by the Mohorovičić discontinuity, or Moho. The mantle extends down to a depth of 2,900 kilometers, where a discontinuous change in physical properties was found. Below this depth is the outer core, shown by the English geophysicist Harold Jeffreys to be fluid rather than solid. At the center is a solid inner core.

The analysis of worldwide seismological observations also revealed that earthquakes occur in a certain pattern over the earth's surface—which had not been suspected on the basis of earlier local observations. In the 1950s Marie Tharp was one of the first to show that the global system of ridges on ocean floors coincided with major zones of earthquake activity. It was already recognized that trenches, mostly on the bottom of the Pacific Ocean near Asia, were sites of frequent earthquakes and volcanic activity. Hugo Benioff, an American geophysicist, found a few regions near trenches where zones of earthquake activity extended several hundred kilometers down below the crust, much deeper than the majority of earthquakes.

In 1960 the American geologist Harry Hess pro-

posed that the earth's mantle, though "solid" in response to rapid deformations, had large-scale patterns of slow convection currents. Material rises through the mid-ocean ridges and spills out, spreading along the ocean floor. At the same time the floor moves away from the ridges toward the trenches, where it descends into the mantle. The Benioff earthquake zones result from this latter process.

The concept of seafloor spreading was soon confirmed in an unexpected way. Magnetic measurements revealed that regions of the ocean floor on either side of ridges show a pattern of alternating stripes of direct and reverse magnetization. Frederick Vine and Drummond Matthews in England analyzed the magnetic data and attributed the pattern to the fact that mantle material, extruded through ridges, cools and acquires a magnetization determined by the earth's magnetic field at the time. A similar proposal was made by Lawrence Morley in Canada. The pattern was found to correspond precisely to the time scale for geomagnetic reversals over the past few million years that was established in the United States.

Hess's hypothesis was developed into a general theory of "plate tectonics" by J. Tuzo Wilson in Canada, along with other scientists in the United States and Europe. According to this theory, the earth's crust and upper mantle down to a depth of about 60 kilometers is composed of relatively strong rigid rock material that is split into six large plates and several smaller ones. As molten material rises from the earth's interior, the plates move, carrying the continents. Plates meet along ocean ridges (where two plates are being pushed apart), trenches (where one dives under another), or transform faults (where one plate is sliding past another, like the San Andreas fault in California). Wherever plates meet, their interaction produces friction and heating, resulting in enhanced earthquake and volcanic activity.

The establishment of plate tectonics in the 1960s is often called a revolution in the earth sciences because it produced radical changes in the theories and methods of geology, geophysics, and oceanography. It is expected to produce substantial practical benefits by explaining why earthquakes occur and thereby allowing them to be predicted.

Stephen G. Brush

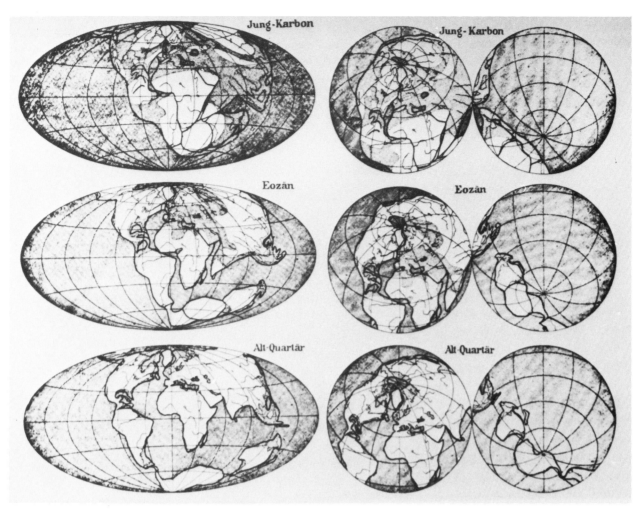

Jung-Karbon

Jung-Karbon

Eozän

Eozän

Alt-Quartär

Alt-Quartär

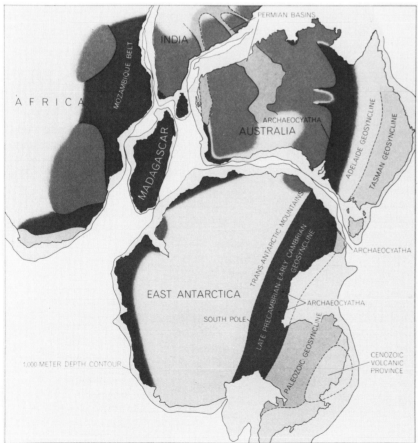

AFRICA

MOZAMBIQUE BELT

INDIA

PERMIAN BASINS

MADAGASCAR

ARCHAEOCYATHA

AUSTRALIA

ADELAIDE GEOSYNCLINE

TASMAN GEOSYNCLINE

ARCHAEOCYATHA

TRANS-ANTARCTIC MOUNTAINS

LATE PRECAMBRIAN-EARLY CAMBRIAN GEOSYNCLINE

EAST ANTARCTICA

SOUTH POLE

ARCHAEOCYATHA

1,000 METER DEPTH CONTOUR

PALEOZOIC GEOSYNCLINE

CENOZOIC VOLCANIC PROVINCE

153. Continents adrift. Although the similarity between coastlines on opposite sides of the Atlantic Ocean was noticed centuries ago, not until 1912–1915 was a comprehensive theory of continental displacement put forth, by the German astronomer and meteorologist Alfred Wegener. He proposed that all the continents were once joined together in a giant landmass. The above maps show how he thought the world appeared in the late Carboniferous, Eocene, and early Quaternary. Not until the 1960s, with new data, was Wegener's controversial hypothesis accepted by the majority of scientists.

154. Gondwanaland. Contrary to Wegener's picture above, later theorists concluded that his protocontinent broke up into supercontinents, such as Gondwana in the southern hemisphere. This map from the late 1960s shows the pieces fitting together at the 1,000-meter (3,280-foot) depth contour of the continental shelves. The ages of the geologic strata match well across Antarctica, Australia, India, and Africa.

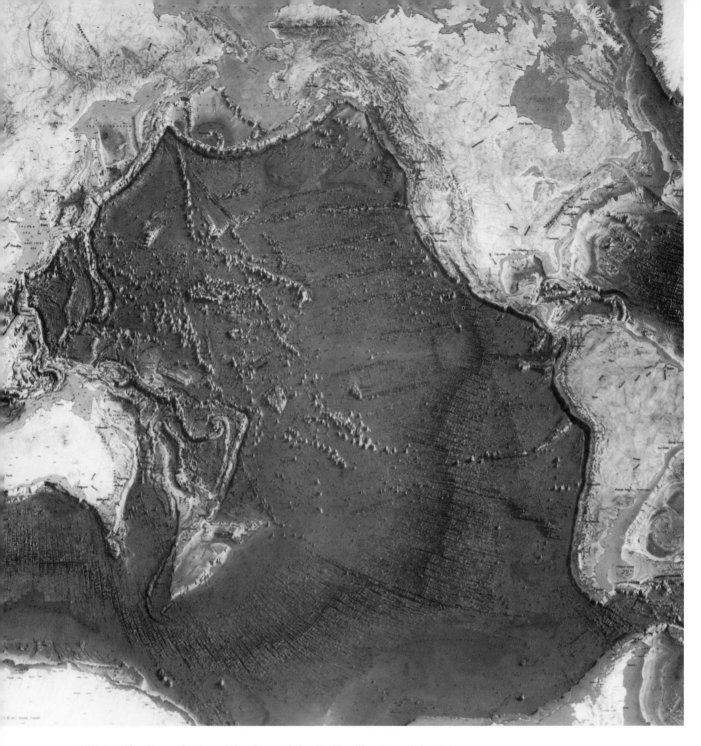

155. Pacific Ocean bottom. The floor of the Pacific, like that of the Atlantic, has its mid-ocean ridges from which new oceanic crust spreads. Note particularly the prominent corrugated structure lying westward and southward of Lower California. Much of the Pacific is relatively uniform in depth, but, as this 1977 topographical map by Bruce Heezen and cartographer Marie Tharp shows, there are deep trenches along the western perimeter (where crustal material is subducted under the continental plates), and there are also scattered volcanic islands and seamounts. Many seamounts, those called guyots, have flat tops, presumably owing to wave erosion of their peaks when they were at the ocean surface. Now some of them lie a mile or more below sea level.

156. Seafloor spreading. A powerful argument for continental drift was provided in the 1960s by the discovery of regions of alternating magnetic polarities on the ocean floor. Symmetrical around the mid-ocean ridge, they were interpreted as a "fossil" record of the earth's changing magnetism. Molten rock welling up along the mid-ocean ridge was magnetized in the direction of the earth's magnetic field as the rock cooled and gradually moved outward; bands of reverse polarity would take shape when the earth's field was reversed.

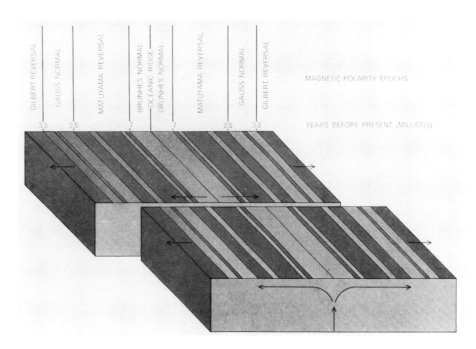

157. San Andreas fault. The fracture in the earth's crust known as the San Andreas fault in California is caused by two tectonic plates—the North American and the North Pacific—sliding past one another. This contact between the two plates accounts for the large, shallow-focus earthquakes the region has experienced, such as the quake that struck San Francisco in 1906. Shown here in a 1968 photograph is the fault in the Carrizo Plain region of southern California.

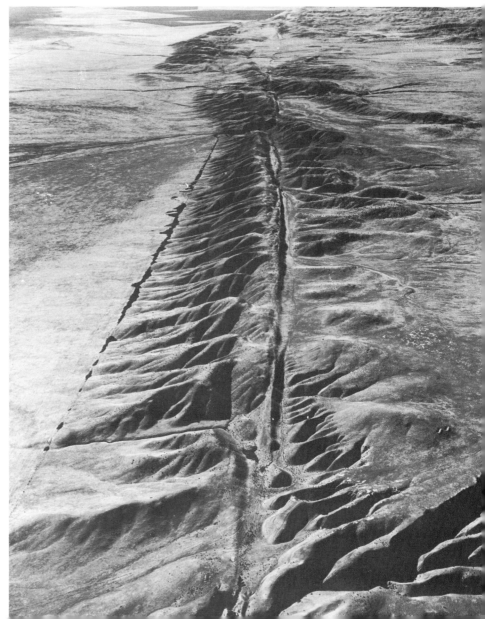

The Daily Mirror

THE MORNING JOURNAL WITH THE SECOND LARGEST NET SALE

No. 3,049.   Registered at the G.P.O. as a Newspaper.   FRIDAY, AUGUST 1, 1913   One Halfpenny.

DEATH OF PROFESSOR JOHN MILNE, THE FAMOUS ENGLISH INVENTOR OF INSTRUMENTS TO DETECT EARTHQUAKES AND REGISTER THEIR MOVEMENTS.

158. The father of seismology. John Milne's research placed studies of earthquakes on a modern basis. He invented the first practical seismograph, he compiled major catalogs of earthquakes, and he pushed for the establishment of seismological monitoring stations throughout the world. From 1875 to 1895 he was professor of geology and mining at the Imperial College of Engineering in Tokyo. His death in 1913 provoked a public response rare for a scientific figure. The front page of this London newspaper shows samples of Milne's seismographic records and a picture of his seismograph, as well as portraits of Milne and his Japanese wife.

159. Agassiz upended. Even the icons of science are not immune to the destructive power of continental motions. This statue of the great nineteenth-century naturalist and geologist Louis Agassiz at Stanford University was overthrown by the earthquake that devastated San Francisco in April 1906. The earthquake resulted from a break along the San Andreas fault that produced horizontal displacements of up to 6 meters (20 feet).

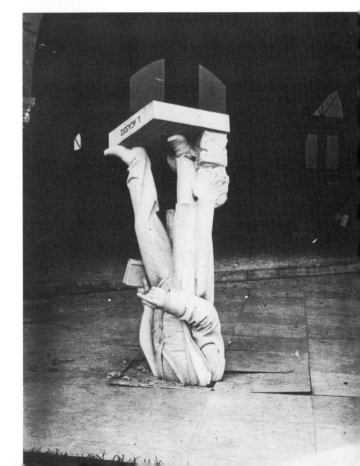

160. San Francisco earthquake. These flats on Golden Gate Avenue collapsed away from the street in the 1906 quake, contrary to the usual tendency during an earthquake. The picture was taken before the block was destroyed by fire. The fire that swept San Francisco after the shock accounted for far more damage than the quake itself.

161. Seismographic monitoring. This photograph was taken at the volcano Mount St. Helens in Washington State in April 1980. After earthquakes and minor eruptive activity began the previous month, geologists increased their monitoring efforts. A major eruption, related to the gradual motions of crustal plates, occurred the following month.

162. Under the volcano. The century opened with a volcanic eruption that was extraordinarily devastating in terms of the death toll. The photograph shows the town of St. Pierre on the Caribbean island of Martinique after the eruption of nearby Mount Pelée on May 8, 1902. The eruption produced a "nuée ardente"—a fiery cloud of ash and other material—that raced down the slope and engulfed the town, killing nearly 30,000 people.

163. Parícutin. One of the very few volcanoes observed from birth, Parícutin erupted in a Mexican cornfield in 1943. The growing cinder cone reached a height of 168 meters (550 feet) in six days and 365 meters in seven months. By the time the eruption ended in 1952, the peak was at 2,800 meters. The photograph was taken in March 1944.

164. Mount St. Helens erupts. This photograph was taken on May 18, 1980, when Mount St. Helens produced one of the greatest volcanic explosions ever recorded in North America. The billowing plume of the eruption extended upwards over 18,000 meters (60,000 feet). The eruption ripped away the uppermost part of the mountain's north flank. The event drew the attention of hundreds of scientists and was history's best documented volcanic display.

165. Birth of an island. In 1963 the volcanic island Surtsey emerged from the Atlantic Ocean south of Iceland, formed by material up-welling at the mid-ocean ridge. Initial emissions of ash, pumice, and steam were followed in April 1964 by lava flows, as Surtsey began to take shape. The three-and-a-half-year eruption was one of the more dramatic evidences of the forces that power seafloor spreading and continental drift.

166. The view from above: north and south. With the advent of artificial earth satellites, meteorology gained a new arsenal of observational capabilities. The cloud patterns in these March 1975 images centering on the north and south poles show the so-called Coriolis effect, the seeming deflection in the path of an object moving on a rotating body like the earth. In the northern hemisphere the cyclonic patterns spiral outward clockwise; in the southern, counterclockwise. The Coriolis forces produce the prevailing westerlies in the temperate zones and the easterly trade winds in the tropics.

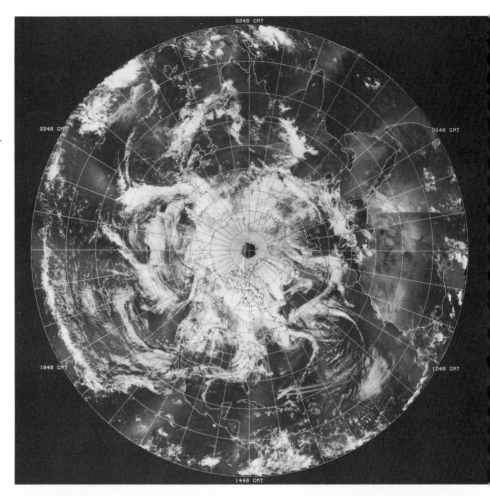

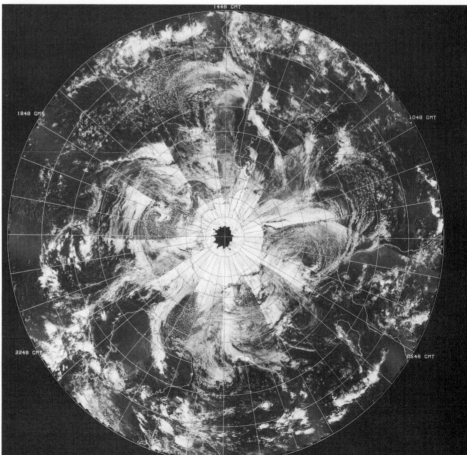

11

Meteorology

If we define meteorology as the art of predicting the weather, then it is as old as mankind. But meteorology could not be called an exact science before the twentieth century, and even after World War II it was still in its infancy. Meteorology as a science was founded on nineteenth-century thermodynamics, radiation theory, and hydrodynamics, but it required a worldwide data-collecting system that was still only rudimentary in the late 1800s.

Physicists such as Hermann von Helmholtz and James Clerk Maxwell initiated the successful marriage of physical methods and meteorological problems in the 1870s and 1880s, but not until later did researchers develop their physical concepts to fit the sort of meteorology for which actual data might be gathered. Such men were Julius von Hann, William Napier Shaw, and Vilhelm Bjerknes.

The Viennese Hann was a leading European meteorologist in the first two decades of the twentieth century. He tried to coordinate empirical and theoretical results in a coherent structure of weather patterns. Convinced that progress was possible only through extensive observations, he urged that upper-air data be collected, for example, at mountain observatories. Shaw, the most prominent British meteorologist of the century, transformed the British Meteorological Office into a modern scientific organization whose staff went beyond the customary statistical correlations to study the connections between physics and the atmosphere.

Hann and Shaw are representative of the organizational and logistical structuring of meteorology in the twentieth century, but the Norwegian Bjerknes is best seen as a theorist. He expanded classical hydrodynamic theory and generalized it, hoping to apply it to the circulation of the earth's atmosphere and oceans. Formulating relationships between air velocity, density, pressure, temperature, and humidity, as well as expressions for the Coriolis forces on a rotating earth, he attempted to solve the equations graphically to establish global weather patterns. More than anyone else, perhaps, Bjerknes represented the trend toward constructing the science of meteorology in the most rigorously mathematical manner, and from his institute in Norway came several of the century's leading meteorological theorists.

Despite the vision displayed by meteorologists in the century's early decades, scientific weather prediction could not become a practical reality until two essential requirements were met. On the one hand, it was necessary to have a more extensive data base, particularly with respect to the upper atmosphere. Secondly, greater computing power was essential; before high-speed digital computers, the task remained hopeless.

Several technological advances helped change the face of meteorological data gathering. Radio, balloon-borne radiosonde probes, and the airplane made possible wider coverage, with respect both to surface area and to height above the earth. Not only did the development of aviation bring technical means of documenting higher atmospheric layers, but the growth of air travel as a medium of commerce transformed the world's meteorological organizations into large bureaucratic structures.

The difficulty of calculating the weather by hand from physical principles is well illustrated in the work done during and after World War I by the English meteorologist Lewis Fry Richardson. Using numerical methods to solve the type of problem that Bjerknes had attacked graphically, Richardson labored for six weeks to advance the weather by a mere six hours—and then found his results completely outside the range of observed pressures. Richardson envisioned 64,000 well-orchestrated human computers trying to outrace the weather. "Perhaps some day in the dim future it will be possible to advance the computations faster than the weather advances and at a cost less than the saving to mankind due to the information gained. But that is a dream," he concluded.

At a conference following World War II, the participants deemed it premature to base weather prediction on scientific principles. Yet, within a decade, the U.S. Weather Bureau was regularly doing just that, largely thanks to the vision of one man, the Hungarian-born mathematical physicist John von Neumann. During World War II von Neumann became interested in the possibilities of digital computers. He recognized weather prediction as a field that could benefit immensely from high-speed electronic computing, especially after prodding from Carl-Gustaf Rossby, a former student of Bjerknes's in Bergen who had been instrumental in bringing American meteorology into world leadership.

In 1948, Jule Charney joined von Neumann's program at Princeton, having worked previously with Rossby. Two years later Charney was able, in an intensive thirty-three-day expedition to the ENIAC computer at the Aberdeen Proving Ground, to produce a twenty-four-hour forecast in twenty-four hours. He showed that such a forecast was better than any that humans could do by ordinary methods, and it seemed likely that if the technique were further developed, the forecast could be made more quickly. By the spring of 1952, using the computer at the Institute of Advanced Study in Princeton, he employed a two-dimensional model (augmented with a layering of the atmosphere) to "retrodict" successfully the Thanksgiving 1950 storm that had caught forecasters unawares and caused $100 million in damage on the U.S. East Coast. By April 1954 a weather prediction unit funded jointly by the U.S. Weather Bureau and the armed services began to compute daily predictions.

A few years later, with the advent of the space age, new possibilities entered the research arsenal of meteorological studies. By the 1960s weather satellites regularly monitored global conditions from above. The temperatures of ocean currents proved to have fundamental effects on climatic patterns, and this, too, could be checked by satellites. Meanwhile, instrument-bearing rockets, radar, radiosonde balloons, and specially equipped airplanes continued to collect data at altitudes lower than those traversed by the satellites.

The dream of predicting the weather far in advance was shattered, however, by mathematical studies in the 1980s. Already foreshadowed in the early 1960s by the findings of the American meteorologist Edward Lorenz, this work showed that minute errors of measurement at the start of a computation could build into major proportions with the passage of time. Lorenz called it the butterfly effect: the flapping wings of a butterfly in Tokyo could have profound consequences for the weather over Europe weeks or months later. Mathematical analyses of this sort generated a new field of study, called chaos, in which seeming disorder appears to be organized into patterns at higher levels. Thus, while details of the weather cannot be ascertained more than a week or so in advance, it may nevertheless be possible to predict broad patterns of meteorological behavior.

Owen Gingerich

168. The 1931 flight. Below we see Piccard's balloon being prepared for launch at Augsburg, Germany. The stratospheric flight ended on an Austrian glacier. Piccard and Kipfer were feared dead; when they were found alive, the acclaim greeting their feat was tremendous. After his last flight, in 1937, Piccard devoted himself to underwater exploration; in 1953 he descended to a record 3,100 meters (10,200 feet), tripling the depth reached by William Beebe and Otis Barton two decades earlier.

167. Ascent to the stratosphere. Auguste Piccard gained worldwide fame when, in May 1931, he and his assistant Paul Kipfer made the first ascent to the stratosphere, reaching an altitude of 15,780 meters (51,800 feet) in a revolutionary balloon with an airtight, pressurized cabin. In August of the following year Piccard, with Max Cosyns, went even higher, to 16,200 meters (53,200 feet). That ascent, from Zurich at night, was reported on in detail by the French picture weekly *L'Illustration*, whose cover is reproduced here.

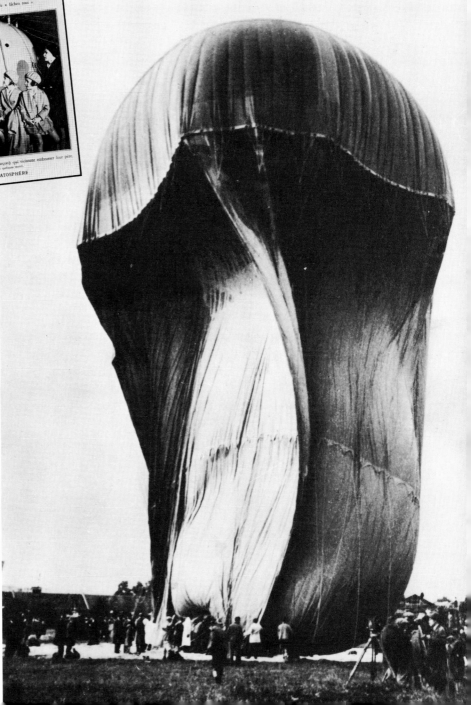

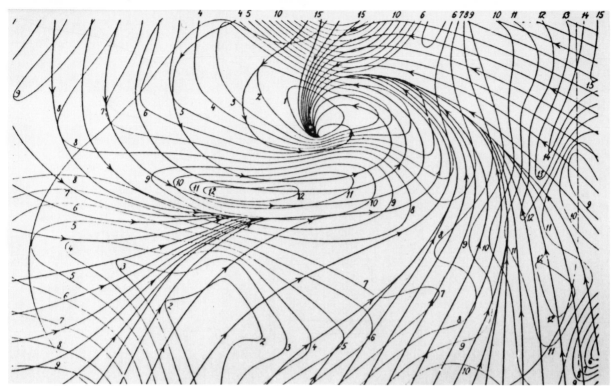

169. Bjerknes and scientific weather forecasting.

Early in this century the Norwegian physicist Vilhelm Bjerknes undertook the immensely complex project of using physical laws to predict future states of the atmosphere. Not until the advent of electronic computers could meteorology even begin to come to grips with the calculations involved. But Bjerknes's failure to achieve his primary goal did not prevent him from making significant advances in practical meteorology. The diagram is from *Dynamic Meteorology and Hydrology*, which Bjerknes coauthored beginning in 1910.

170. The Bergen school.

The influential school of meteorology founded in Bergen by Bjerknes was housed from 1919 in the upstairs rooms of his large frame house. Below, in one of the rooms, scientists Tor Bergeron (*center*) and Jacob Bjerknes (*right*) are analyzing meteorological data. At left, a woman is recording information on weather maps. Bergeron developed a widely accepted system for classifying air masses. Jacob Bjerknes, Vilhelm's son, discovered that low-pressure centers, or cyclones, begin as waves in sloping fronts dividing different air masses.

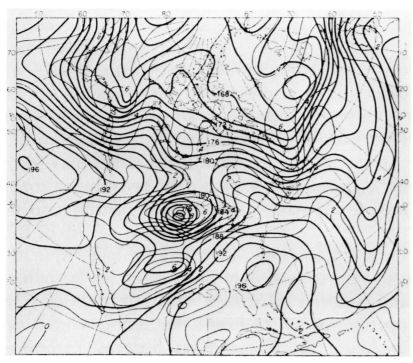

171. Meteorology comes of age.
The maturation of meteorology as a science was largely due to the Swede Carl-Gustaf Rossby, who worked briefly at Bergen under Vilhelm Bjerknes but who spent his most productive years (1926–1950) in the United States. Rossby introduced fundamental equations describing large-scale atmospheric motion. Above, he stands beside a rotating tank built for him by the U.S. Weather Bureau in 1926–1927 for experiments on fluid motions.

172. Computer weather prediction.
The American meteorologist Jule Charney developed a mathematical theory of atmospheric motions that was applied in 1950 with some success to yield a numerical prediction of weather patterns, made possible by using the newly developed electronic computer. Here is one of the maps resulting from this pioneering experiment at the Institute for Advanced Study in Princeton, New Jersey.

173. Kite sounding. In the nineteenth century manned balloon flights acquired much interesting meteorological information, but it was often fragmentary and inaccurate. A more reliable method, the use of kites, was widely adopted around 1893, and by 1898 the U.S. Weather Bureau operated sixteen kite stations. The photograph shows a kite sounding in 1900 in Arlington, Virginia. The kite meteorograph was capable of recording temperature, pressure, and humidity. Meteorological kite flying continued into the 1930s, when it was replaced by the airplane.

174. Pilot balloon. In this World War I scene, two German soldiers are preparing a hydrogen-filled balloon for launching from a field station. By following the ascent of the small pilot balloon with the theodolite at left, the speed and direction of the wind aloft could be estimated. Balloons could penetrate much higher than kites; it was with balloons that the stratosphere was discovered at the turn of the century. Balloon-borne devices for measuring temperature, humidity, and air pressure and then radioing the data back to the surface came into use in the 1930s.

175. Sampling air aloft. In addition to a meteorograph, the balloon in this picture will carry aloft a glass container for obtaining a high-altitude air sample. The device was designed by the Austrian-born chemist Friedrich Paneth, who, beginning in the mid-1930s, studied the trace components of the atmosphere. Paneth determined the content of helium, ozone, and nitrogen dioxide in the atmosphere and concluded that the components are not gravitationally separated until an altitude of about 65 kilometers (40 miles).

176. Hurricane hunters. Toward the end of World War II the U.S. military began air reconnaissance of hurricanes and typhoons, having found that it was possible to fly into a hurricane and return alive. Hurricane hunting with specially equipped planes became a regular practice. The photograph shows a group of U.S. Air Force hurricane hunters preparing for a reconnaissance flight from Bermuda in 1952. Air surveillance of storms was still in its infancy, and the mission was a dangerous one. Quite apropos was the squadron's emblem visible on the plane: a crystal ball guides an airman under his umbrella through a storm on a flying carpet.

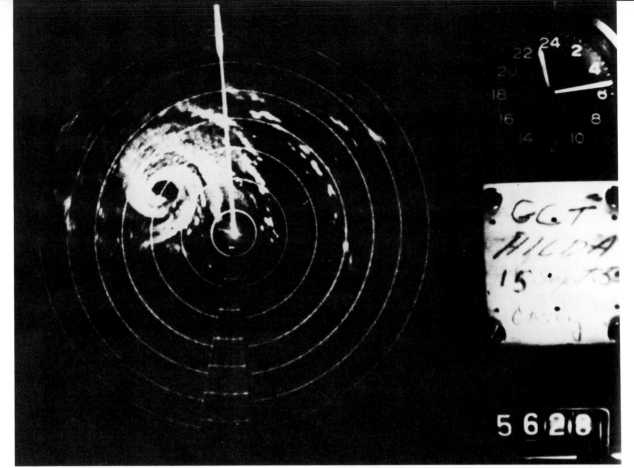

177. Tracking hurricanes by radar. The above radar image of Hurricane Hilda, which struck Mexico in September 1955, was captured by an airborne early-warning plane. Weather radar, first used on storm-hunting aircraft in 1945, informed meteorologists that heavy precipitation was concentrated in spiral arms issuing from the eye of the storm. The radar waves bounce back from the water and ice particles in the atmosphere, becoming visible as the bright areas on the cathode-ray screen.

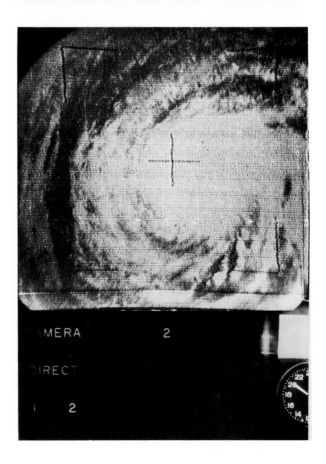

178. A hurricane detected from space. Esther was the first hurricane or typhoon to be discovered by satellite imaging. The first weather satellite, TIROS 1 (for Television and Infrared Observation Satellite), was launched into orbit in April 1960 and quickly imaged a typhoon off Australia. But it was TIROS 3 that in September 1961 found a severe cyclonic storm over the Atlantic whose existence had not yet been suspected by earthbound observers: Hurricane Esther. The fearsome Esther seemed to be heading for Virginia but in the end remained over the ocean. Nevertheless, the satellite's early warning provided enough time to carry out a silver iodide cloud-seeding experiment aimed at reducing Esther's intensity; the trial, however, was inconclusive.

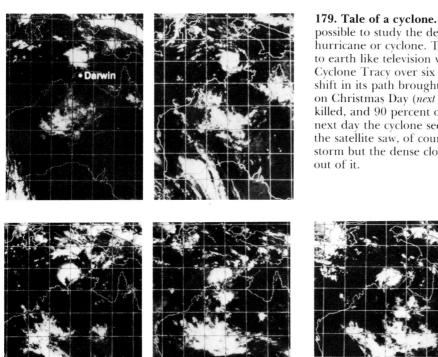

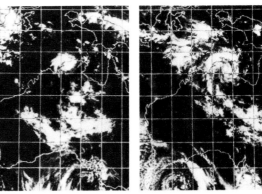

179. Tale of a cyclone. Satellite observation made it possible to study the development and progress of a hurricane or cyclone. These pictures, transmitted back to earth like television views, show the erratic course of Cyclone Tracy over six days in late 1974. A sudden shift in its path brought disaster to Darwin, Australia, on Christmas Day (*next to last picture*); 50 people were killed, and 90 percent of the city was destroyed. By the next day the cyclone seemed to be dissipating. What the satellite saw, of course, was not the heart of the storm but the dense clouds lying above and spiraling out of it.

180. Cloud seeding. The end turn of this racetrack in the sky is 5 miles, or 8 kilometers, long; the straightaway, 20 miles. The pattern was the result of a successful 1947 experiment in weather modification. A supercooled stratus cloud was seeded with crushed dry ice, causing the supercooled water droplets in the cloud to form ice crystals that led to a snow shower. The racetrack is the path followed by the aircraft in seeding. The 1940s saw an upsurge of interest in weather modification after it was discovered that dry ice (1946) and silver iodide (1947) could help supercooled water droplets form tiny ice crystals.

Part Four

HARNESSING
THE ATOM

181. Millikan in search of cosmic rays. Robert Millikan, seen here with a high-altitude balloon in 1938, began studying cosmic radiation in the 1920s. He showed that the radiation, discovered earlier by the Austrian physicist Victor Franz Hess, indeed had an extraterrestrial origin. Millikan coined the name "cosmic rays" in 1925.

12

Nuclear Physics

After the invention of quantum mechanics in 1925–1926, it appeared for a short time that a complete description of the physical world could be based on only three fundamental entities, each having both wave and particle properties: the proton, the electron, and the photon (electromagnetic radiation). Aside from gravity, which seemed to be of negligible importance on the atomic scale, the only fundamental forces known in 1926 were electricity and magnetism.

When P. A. M. Dirac, a mathematical physicist at Cambridge University, tried to make quantum mechanics consistent with Einstein's special theory of relativity, he found that he could derive a satisfactory equation for the electron's wave function if that function had four components. Two components corresponded to the two different states of spin postulated by Uhlenbeck and Goudsmit, but the other two apparently belonged to a new particle, similar to the electron but with positive electric charge.

In 1932 the American physicist Carl D. Anderson discovered tracks of positively charged particles on cloud-chamber photographs of cosmic rays. These particles, which he called positrons, were soon recognized to be Dirac's predicted counterpart to the electron.

A second important discovery in 1932 was made by the British physicist James Chadwick. Rutherford had earlier predicted the existence of the neutron, a particle with about the same mass as the proton but no electric charge. Chadwick identified the neutron as a particle formed when beryllium is bombarded with alpha particles and confirmed that its mass is close to that of the proton.

The neutron immediately proved very useful in understanding the composition of the atomic nucleus. Before 1932 physicists assumed that the nucleus is made of protons and electrons, but this theory was unsatisfactory for several reasons. Werner Heisenberg proposed instead that the nucleus is composed of protons and neutrons. Thus the nuclear charge (equal to the atomic number) is just the number of protons; the nuclear mass number (the integer closest to the atomic weight of the particular isotope) is the total number of protons plus neutrons.

A third new particle appeared in the physics literature in 1932, though it was not actually discovered in the laboratory until the 1950s. The neutrino was postulated by Wolfgang Pauli in order to account for the missing mass and energy in beta decay; Enrico Fermi named it and worked out a detailed theory of its behavior.

Fermi's theory of beta decay involved a new force, now called the weak force. Also, it was assumed that protons and neutrons in the nucleus were held together by another kind of force, the "strong" force effective only at very small distances. In this way, physicists came to believe that four fundamental forces exist in nature: gravity, electromagnetism, the weak force, and the strong force.

There is a lingering prejudice in physics against action at a distance—the idea that one object can exert a force on another through empty space, without any contact or material substance to transmit the force between them. One can always try to eliminate action at a distance by postulating that a force is transmitted by particles. Thus, in the quantum field theory developed in the 1930s, electrical forces were described in terms of the continual emission and absorption of photons.

In 1935 the Japanese physicist Hideki Yukawa extended this idea by attributing the strong force to the exchange of a hypothetical particle between protons and neutrons. Because of the short-range character of the force, the new particle had to have a finite mass, estimated to be about 200 times that of the electron.

Yukawa's prediction became known to Western physicists in 1937, about the same time as the discovery of the muon. It was thought for several years that the muon was the particle postulated by Yukawa, since it had approximately the same mass. But an experiment in 1947 showed the interaction between muons and atomic nuclei to be very weak, indicating that muons could not be the carrier of the strong force. At about the same time the pion was discovered. It interacted strongly with nuclei and was therefore identified as Yukawa's particle.

Most of these particles were discovered in cosmic rays, but during the 1940s and 1950s the use of accelerators such as the cyclotron took on increasing importance in nuclear and particle physics. The American physicist Ernest Orlando Lawrence became the leader in constructing accelerators to produce particles with ever-higher energies.

Research on the properties of the nucleus advanced along with discoveries of new particles. Although the nucleus, like the atom, consists of particles held together by attractive forces, its structure is more difficult to understand because no central particle dominates the others in the way that the nucleus dominates the electrons in the atom. Nevertheless physicists tried to understand how protons and neutrons might be arranged in a way analogous to the "shells" of electrons within the atom. A successful shell model of the nucleus was developed in 1950 by Maria Goeppert Mayer in the United States and researchers in Germany.

While some physicists were trying to understand what holds the nucleus together, others were finding out how to take it apart. In 1934 an important experiment was conducted in France by Irène and Frédéric Joliot-Curie. (Irène was the daughter of Marie Curie.) They found that when certain light elements (boron, magnesium, aluminum) were bombarded with alpha particles from polonium, positrons as well as protons and neutrons were emitted, and positrons continued to be emitted after the alpha source was removed. It appeared that an initially stable nucleus had been changed into a radioactive one.

Enrico Fermi and his colleagues in Italy then undertook a systematic study of nuclear reactions induced by neutrons, looking for new elements that might be produced in this way. Fermi thought that by bombarding uranium, the heaviest known element, with neutrons, an even heavier, transuranium element might be produced. In fact he did produce a radioactive element, but in such a small amount that he could not determine what it was. Ida Noddack, a German chemist, suggested that Fermi's experiments, instead of producing a heavier element, caused the nucleus to break into several large fragments. Her suggestion, however, was ignored.

The discovery of nuclear fission was accomplished by a group consisting of two German scientists—Otto Hahn and Fritz Strassmann—and two Austrians, Lise Meitner and her nephew Otto Frisch. In 1938, Hahn and Strassmann showed that one of the substances that Fermi thought might be a transuranium element was actually an isotope of barium; they also found that isotopes of lanthanum, strontium, yttrium, krypton, and xenon were produced. They were reluctant to accept what now seems the obvious conclusion, that the nucleus of uranium had split into two smaller nuclei. This conclusion was soon proposed by Meitner and Frisch. They introduced the term "fission" and noted that enormous amounts of energy could be released by the transformation of a small amount of the mass of the uranium nucleus.

Scientists in several countries immediately started to study nuclear fission. It was quickly realized, by the Hungarian-born physicist Leo Szilard and others, that the fission reaction produces neutrons that could initiate other fission reactions, resulting in a self-sustaining chain reaction. In 1939, as Europe prepared for another world war, the physicists began to consider the possibility that nuclear fission could lead to a new weapon, so powerful that it might decide the outcome of that war.

Stephen G. Brush

182. Hess in his gondola. Although previous scientists had observed an enigmatic, extremely penetrating radiation at the earth's surface, Hess may truly be regarded as the discoverer of cosmic rays. In a series of daring balloon ascents in 1911 and 1912, he proved that this radiation increases markedly in intensity with altitude. He also showed that the intensity is the same both day and night. Hess concluded that the radiation was of extraterrestrial, but not solar, origin. The photograph was probably taken during the landing of a test flight in 1912.

183. The first antiparticle. The Wilson cloud chamber photograph at left, made by the American physicist Carl D. Anderson in 1932, shook the world of physics. The characteristics of the curved track were such that only a particle of the same mass as the electron but with positive charge could have produced it. The mathematical possibility of positive electrons, or positrons, had been shown by the British physicist Paul Dirac, but there had not been any evidence that they actually existed. Below is Anderson with the positron discovery apparatus.

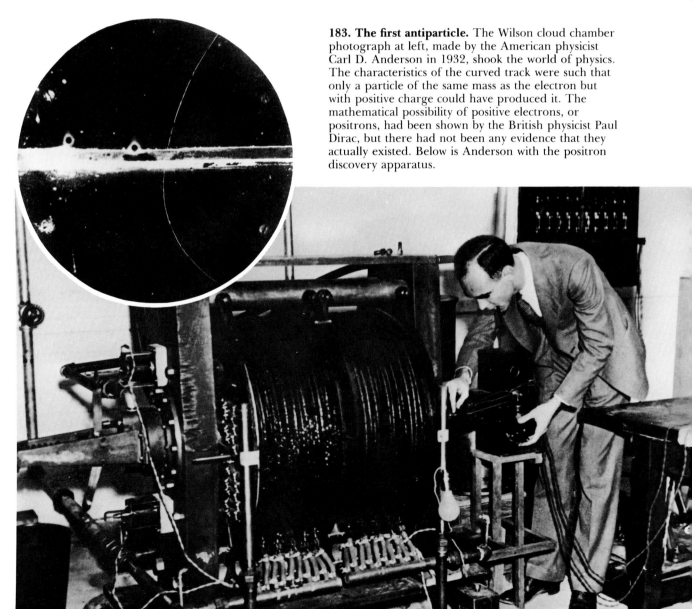

186. Fermi and Bohr. Here the ubiquitous Niels Bohr engages in conversation with the Italian Enrico Fermi during a walk along the Appian Way. The year was 1931. The young Fermi was already famous as an atomic physicist and had accepted the first Italian chair of theoretical physics, at the University of Rome. He was soon to make important discoveries in experimental nuclear physics. Fermi's unusual versatility made him a rara avis, excelling both as a theoretician and experimentalist. ▶

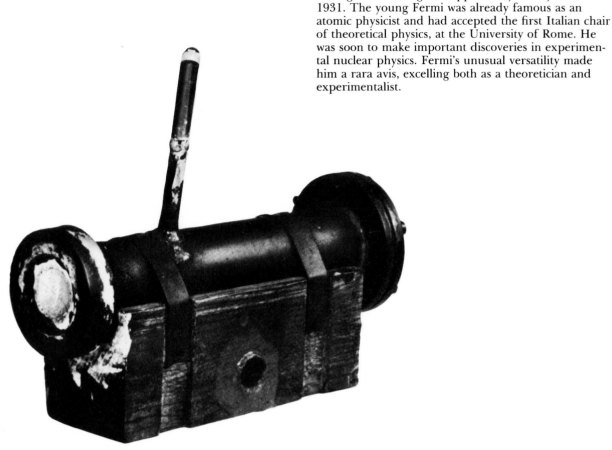

184. Chadwick's neutron apparatus. With this device containing a radioactive source and a counter, James Chadwick discovered the neutron. Physicists like Rutherford had suspected that such a particle existed in the nucleus, perhaps as a paired electron and proton. But in 1932 at the University of Cambridge's Cavendish Laboratory, Chadwick became the first to demonstrate that the neutron existed as an independent entity.

185. Inside the apparatus. The two basic parts were a polonium-beryllium source (*left*) and an ionization chamber counter (*right*). Alpha particles from the polonium struck the beryllium, producing a "radiation" that had been observed by other scientists but whose nature was unknown. Chadwick put a sheet of paraffin between the source and the ionization chamber; after measuring the nuclei ejected from the sheet by the "radiation," he concluded that it was the long-sought neutron.

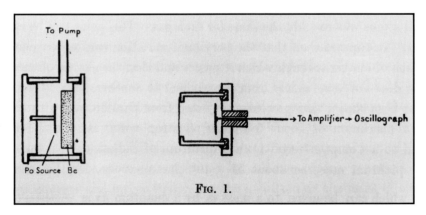

To Pump

Po Source   Be

→ To Amplifier → Oscillograph

FIG. 1.

A Consistent Theory of the Nuclear Force
and the β -Disintegration

In spite of many attempts to develop the so-called "β-hypothesis of the nuclear force",[1] there still remains in the current theory the well known inconsistency between the small probability of the β-decay and the large interaction of the neutron and the proton. Hence, it will not be useless to give on this occasion a brief account of one possible way of solving this difficulty which was proposed by the present writer about two years ago.[2]

First, we introduce the field which is responsible for the short range force between the neutron and the proton and assume it to be something different from the so-called "electron-neutrino field" in contradistinction to the current theory. The simplest conceivable one is perhaps such that can be derived from two scalar potentials U and Ũ, which are conjugate complex to eac each other and satisfy, in the presence of a heavy particle, the th equations

$$\left\{ \Delta - \frac{1}{c^2}\frac{\partial^2}{\partial t^2} - \lambda^2 \right\} U = -4\pi g \tilde{v} u \qquad \curvearrowright \ 0 \qquad (1)$$

$$\left\{ \Delta - \frac{1}{c^2}\frac{\partial^2}{\partial t^2} - \lambda^2 \right\} \tilde{U} = 0 \qquad \curvearrowright \ -4\pi g \tilde{u} v \qquad (2)$$

187. A prophet spurned: the meson and cosmic rays. In 1935 the Japanese physicist Hideki Yukawa suggested that the force binding protons and neutrons in an atomic nucleus was carried by energy-particles that had not yet been observed. Although his hypothesized particle was too energetic to be made by the atom smashers of the day, the discovery in 1936 of a new particle in cosmic rays led Yukawa to propose, in this letter to the British journal *Nature,* that it was the one he had described. But the editor apparently thought the whole idea too strange to publish. The observed particles turned out to be closely related to the particles Yukawa had predicted. They are all now called mesons, meaning "middle ones," because they have a mass-energy intermediate between that of the electron and proton. Yukawa won a Nobel Prize in 1949.

188. Cockcroft's accelerator. The era of Big Science was unfolding in the Cavendish Laboratory in 1932 as John Cockcroft and Ernest Walton designed this electrostatic accelerator that could speed protons to energies of hundreds of thousands of volts. Cockcroft can be seen in the counting chamber under the discharge tube. With this machine, Cockcroft and Walton achieved the first nuclear disintegrations to be caused by artificially accelerated particles. Their high-energy protons were directed at a lithium target. The ensuing reaction between protons and lithium nuclei resulted in the formation of two alpha particles: the lithium nucleus had vanished.

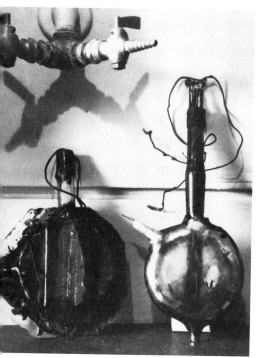

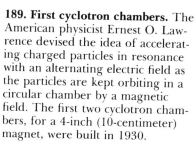

189. First cyclotron chambers. The American physicist Ernest O. Lawrence devised the idea of accelerating charged particles in resonance with an alternating electric field as the particles are kept orbiting in a circular chamber by a magnetic field. The first two cyclotron chambers, for a 4-inch (10-centimeter) magnet, were built in 1930.

190. Lawrence. By the end of 1936 there were a score of cyclotrons in the world. Lawrence himself continued to build new machines at the University of California at Berkeley. Here we see him at the magnet of Berkeley's 37-inch cyclotron, made from an existing 27-inch machine in 1936.

191. 60-inch cyclotron. By 1939, Lawrence's Radiation Laboratory had a cyclotron whose magnet pole faces were 60 inches in diameter. The elements neptunium and plutonium were discovered with it in 1940 after being produced by bombardment of uranium. The machine became an important source of isotopes for research.

192. Meitner and Hahn. These two key figures in the discovery of nuclear fission, the Austrian-born physicist Lise Meitner and the German chemist Otto Hahn, worked together for nearly thirty years. In 1912 they joined the newly opened Kaiser Wilhelm Institute for Chemistry in Berlin, where this 1913 photo was taken. Since the laboratory was new, it was still uncontaminated and permitted the study of weakly radioactive substances.

Atom Explosion Frees 200,000,000 Volts; New Physics Phenomenon Credited to Hahn

By The Associated Press.

WASHINGTON, Jan. 28.—American scientists heard today of a new phenomenon in physics—explosion of atoms with a discharge of 200,000,000 volts of energy.

Theoretical physicists attending a meeting sponsored by the Carnegie Institution of Washington and George Washington University said that Dr. Enrico Fermi of the University of Rome told yesterday that this had been accomplished by Dr. G. Hahn of Berlin.

The report so stirred the limited circle of scientists with facilities to carry on such experiments that work on attempts to duplicate Dr. Hahn's accomplishment has begun at the Carnegie institution's terrestrial magnetism laboratory and at Columbia University.

Scientists at the meeting said the discovery was comparable in significance to the original discovery of radioactivity thirty years ago. They said that it was too soon to discuss possible applications of the

new 200,000,000-volt force, which is thirty times more powerful than radium, but pointed to the fact that radium is now the most efficient weapon used for the treatment of cancer. Like radium, it may be twenty or twenty-five years before the phenomenon could be put to practical use and it might not be practical at all, they said.

Dr. Fermi related that Dr. Hahn bombarded a synthetic element known as "ekauranium" with neutrons, the slow-moving particles of the atom, and produced barium, the substance used in making X-ray pictures of the stomach and intestines.

The only way that this could occur, according to physicists, would be for the ekauranium atom to split apart to form barium and the rare element masyrium.

In causing such a split a force of 200,000,000 volts would be generated since atoms are held together by electrical forces many hundred times more powerful than the force of gravity which holds the stars, planets, sun, earth and moon in their orbits.

PRESIDENT'S GUESTS FOR BIRTHDAY DINNER

man of New York; Kirke Simpson, Charles McCarthy and James P. Sullivan.

The President w— —ears o'

(left margin fragments)
YORK
TOMB

Get Merit
remony
rrow

SCHEDULE

mpletion in
Be Ready
nth

accomplished
res in renovat-
Riverside Drive
vill be paid at
en Lieut. Col.
WPA Adminis-
certificates of
e workers.
being made on
erbert L. Satter-
he Grant Monu-
who will take
ses. Also present
ysses S. Grant 3d
Baker, vice presi-
anization; Brig.
Bates, secretary;
easurer; William
specia'

(right margin fragments)
REPUB'
BA

Westche
Calls A
to E

LA GUAR

Plank
Offici
Pa

WHI
—Leg
legal
York
platf
publi
Cour
gani
tion
In
prop
tione
ther
of
niz

---

193. Fission discovered. Late in 1938, Hahn, working with the physical chemist Fritz Strassmann, found that uranium atoms bombarded with neutrons yielded what appeared to be much lighter barium atoms. Hahn wrote of this discovery to Meitner in Sweden, where she had fled from the Nazis. Meitner and her nephew Otto Frisch realized that the uranium nucleus had been split, and they called the process fission. They noticed that the energy released— 200,000,000 electron volts— matched the mass loss of the uranium. Frisch shared the news with Bohr, who was going to the United States to speak at a conference early in 1939. Reproduced here is a confused press report of the announcement of Hahn's discovery. Scientists were greatly excited but did not openly mention its potential for atomic weapons.

194. Hahn's worktable. This laboratory bench, now preserved at the Deutsches Museum in Munich, holds the equipment with which Hahn split the uranium atom in 1938. Hahn and Strassmann could hardly believe what had happened, "so contrary to all existing experience of nuclear physics. After all, a series of strange coincidences may, perhaps, have feigned these results."

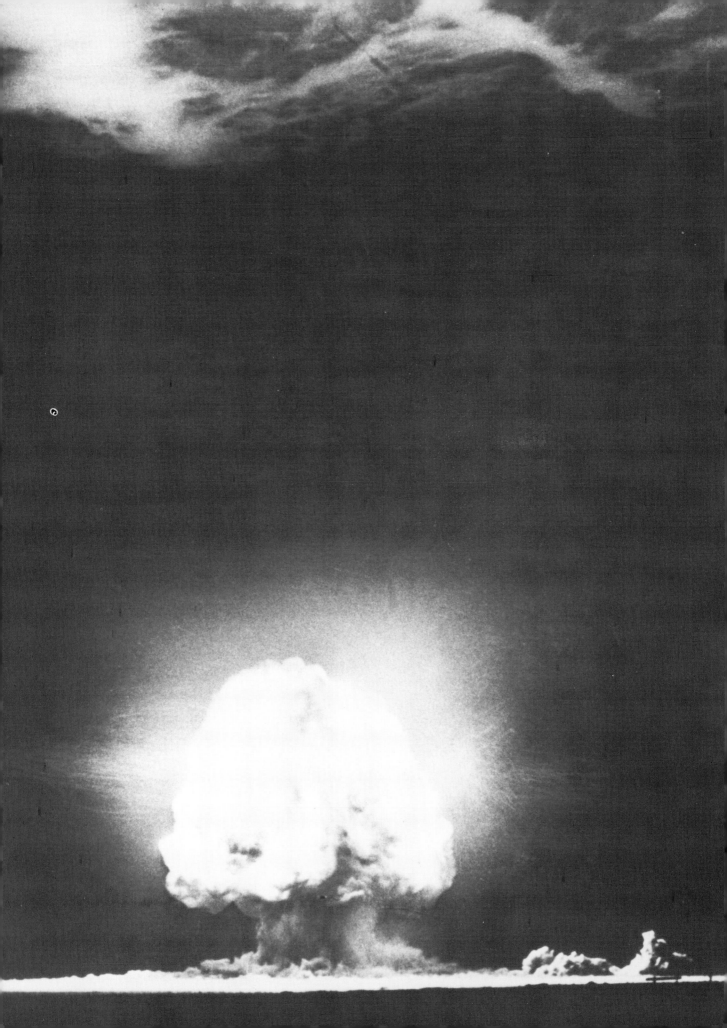

13

Nuclear Fission and Nuclear Fusion

In August 1939, Albert Einstein sent a letter (probably drafted by Leo Szilard) to President Franklin D. Roosevelt, pointing out that "it may become possible to set up a nuclear chain reaction in a large mass of uranium" and that, using such reactions, "extremely powerful bombs of a new type" might be constructed. Although this letter did lead to a small-scale project in the United States, the prospects for actually producing a weapon in time for use in the coming war seemed so remote that little effort was devoted to it.

Two weeks after the outbreak of World War II, Adolf Hitler boasted of powerful new weapons. This alarmed British government officials, who had heard rumors about nuclear explosives, and they asked two German physicists living in Birmingham, Otto Frisch and Rudolf Peierls, for a report. Frisch and Peierls assumed that only the rare isotope uranium 235 was capable of fission; they concluded that a bomb could be made if it were possible to obtain this isotope in nearly pure

195. Trinity test. The first atom bomb explosion took place at the Trinity test site in New Mexico early in the morning of July 16, 1945. After a few seconds this white-hot fireball, photographed from a distance of 6 miles, was rising from the desert. In the blinding light, Los Alamos lab director J. Robert Oppenheimer thought of the Bhagavad-Gita: "I am become Death,/ The shatterer of worlds."

form and if enough of it—the critical mass—could be brought together to sustain a chain reaction. Another possibility, suggested by American and French physicists, was to use element 94 (later called plutonium), to be produced by bombarding the more abundant uranium isotope 238 with neutrons.

When the United States government was informed of the work being done in England, President Roosevelt agreed to revive the American project. In 1942, Colonel Leslie R. Groves of the Army Corps of Engineers was put in charge of the Manhattan Project to develop the atomic bomb.

In order to make a fission bomb, it was necessary either to separate uranium 235 from the other uranium isotopes or to use isotope 238 to produce plutonium. Three different methods were proposed to separate uranium 235: gaseous diffusion, thermal diffusion, and electromagnetic separation. All three were tried and were to some extent successful. At the same time, a group led by Enrico Fermi at the University of Chicago pursued efforts to produce plutonium from uranium 238. They succeeded in obtaining the first controlled nuclear chain reaction in their "atomic pile" on December 2, 1942.

In the spring of 1943 a new laboratory to design the atomic bomb was set up at a remote site in Los Alamos, New Mexico, under the direction of the American physicist J. Robert Oppenheimer. The Los Alamos group considered two possible methods for assembling a critical mass of fissionable material. The simpler was a "gun" that would fire one subcritical mass into another, thus suddenly producing a critical mass. The more complex method involved implosion: a subcritical mass was surrounded with high explosives, deployed in such a way as to produce, when detonated, a converging shock wave that would compress the mass into a small volume; this would have the same effect as exceeding the critical mass.

To be sure that a workable bomb could be produced, the Los Alamos group developed both types: a gun weapon, using uranium 235, and an implosion weapon, using plutonium. The implosion device was tested at Alamogordo, New Mexico, on July 16, 1945. Two fission bombs were then used against Japan. The first, dropped on Hiroshima on August 6, 1945, was a gun-type uranium bomb called Little Boy. The second, dropped on Nagasaki on August 9, 1945, was an implosion-type plutonium weapon called Fat Man.

How did the scientists involved in the development of the atomic bomb justify putting such a terrible weapon at the disposal of politicians, and to what extent did they try to influence the actual use of this weapon? How did the atomic bomb influence the outcome of the war and subsequent events? These are questions still debated by historians. Many scientists, especially among those who had escaped from the fascist countries, feared that Nazi Germany would get the atomic bomb first and use it to win the war. As it turned out, the Germans did not come close to developing a usable bomb, but the atomic scientists in the United States didn't know this until after the war.

Before the defeat of Germany, scientists and policy makers in the United States and Britain assumed that the atomic bomb would be used against Germany if possible; the question of deliberately not using such a weapon apparently was never even discussed. The major issue was whether to keep an Anglo-American monopoly on the bomb after the war in order to discourage Soviet expansionism. But Germany surrendered on May 7, 1945, before the atomic bomb was ready. The question then arose: should the bomb be used against the Japanese? There was no reason to believe that the Japanese had nuclear weapons (although it was later discovered that they did try to develop them). But the Americans were convinced that the Japanese would not surrender unless their homeland were successfully invaded, and this would involve enormous loss of life on both sides. After Roosevelt's death on April 12, 1945, Harry S. Truman became president and made the decision to use the atomic bomb against Japan. The Japanese surrendered on August 14, three days after Nagasaki was destroyed.

The postwar years saw a nuclear arms race unfold between the United States and the Soviet Union. The Soviets tested their first fission bomb in September 1949. Physicists in the United States now intensified their efforts to develop a new kind of atomic bomb, based on the fusion of nuclei. Hans Bethe had shown a decade earlier that the fusion of hydrogen nuclei (protons) to form helium nuclei, together with the formation of heavier elements, might account for the generation of energy in stars. The Hungarian-born American physicist Edward Teller was the chief advocate for developing a hydrogen bomb, using a fission bomb to provide the high temperatures and high pressures needed to initiate the fusion of hydrogen atoms.

Despite objections from Oppenheimer, Teller persuaded the U.S. government to support the development of a bomb based on his ideas. An experimental thermonuclear device with destructive power equivalent to several million tons of TNT was successfully tested in 1952. After a second test, involving a 15-megaton bomb, in 1954, radioactive rain fell on a small Japanese fishing boat in the area, killing one fisherman and poisoning others. Suddenly the world became aware of the continuing danger from radioactive fallout produced by nuclear tests.

Although the destructive uses of nuclear reactions have attracted the greatest public attention, peaceful applications have become increasingly important in science and industry. Scientists employ radioactive isotopes as tracers to study chemical reactions and biological processes; physicians treat cancer and other diseases with radiation. The development of nuclear reactors to produce electric power has bogged down in controversy over safety problems, but nuclear power may be inevitable, since fossil fuels will someday be exhausted. Fusion still promises to provide clean, abundant energy, although the achievement of controlled thermonuclear reactions has proved technically much more difficult than was once believed. We are still trying to learn how to live with the atomic nucleus and its great potential for benefit as well as disaster.

Stephen G. Brush

196. Birth of the Atomic Age. This painting portrays the historic occasion in 1942 when the world's first working nuclear reactor, built in a squash court under the stands of Stagg Field at the University of Chicago, became operational. It was a pile of solid graphite blocks alternating with graphite embedded with uranium. At the left of the pile a structure holds the cadmium control rods. Enrico Fermi, who directed the pile's construction, is standing in front of the instrument rack, slide rule in hand, computing the rise in the neutron count. Above, at right, squats the "suicide squad," poised to dowse the pile with liquid cadmium in case it went out of control.

197. Birth certificate of the Atomic Age. The activity of the Chicago pile was recorded by this galvanometer graph of neutron intensity. The first self-sustaining chain reaction, initiating the controlled release of nuclear energy, took place between three and four o'clock on December 2, 1942. Afterward, Arthur Compton, one of the observers, phoned James B. Conant, the head of the National Defense Research Committee, with a carefully coded message: "The Italian navigator has just landed in the New World." "Already?" asked the surprised Conant. "Yes, the earth was smaller than estimated," replied Compton, "and the natives were friendly."

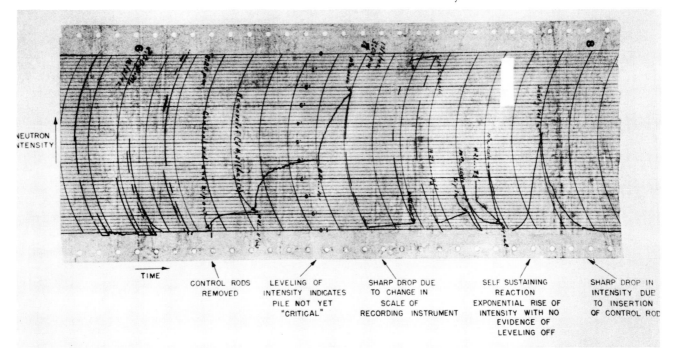

NEUTRON
INTENSITY

TIME

CONTROL RODS
REMOVED

LEVELING OF
INTENSITY INDICATES
PILE NOT YET
"CRITICAL"

SHARP DROP DUE
TO CHANGE IN
SCALE OF
RECORDING INSTRUMENT

SELF SUSTAINING
REACTION
EXPONENTIAL RISE OF
INTENSITY WITH NO
EVIDENCE OF
LEVELING OFF

SHARP DROP IN
INTENSITY DUE
TO INSERTION
OF CONTROL ROD

198. Delivering the test bomb. Detonation of the first atom bomb took place atop a 100-foot (30-meter) steel tower at the Trinity test site. On July 13, 1945, Los Alamos scientists unloaded the device at the tower, where assembly was completed by insertion of the plutonium core and neutron initiator source.

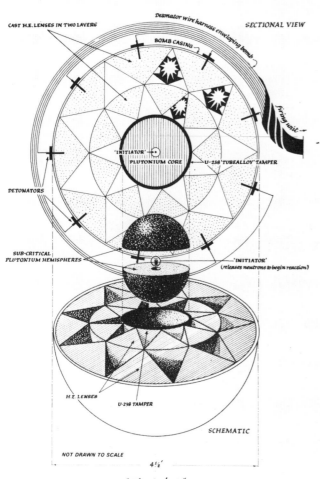

"The Gadget"

199. Inside the gadget. S-1 was the official name of the test bomb that was exploded on July 16, 1945, but most of the scientists who worked on it called it "the gadget." As this artist's schematic shows, the bomb was an implosion device. A charge from the firing unit at upper right detonated a conventional explosive, compressing the core of the bomb and causing the two hemispheres of plutonium 239 to form a critical mass. In fact, the compressibility of the plutonium was essential to get the explosive material compact enough for the chain reaction to begin.

200. Ground zero two months later. Oppenheimer inspected the site with General Leslie R. Groves, who headed the Manhattan Project, the secret U.S. government program that developed the atom bomb. The Trinity explosion had vaporized the steel tower and concrete base and fused the desert sand into a crust. Groves was curiously unimpressed and asked, "Is this all?"

201. Little Boy. The first nuclear weapon ever detonated was of the type code-named Little Boy; it was dropped on Hiroshima, Japan, on August 6, 1945. Unlike the gadget, Little Boy, which weighed 7,000 pounds (3,000 kilograms), was a "gun-type" uranium device. A conventional explosive drove a small piece of uranium into a larger piece, forming a critical mass.

202. Fat Man. This was the type of bomb dropped by a U.S. bomber on Nagasaki, Japan, on August 9, 1945. Like the gadget, Fat Man was an implosion device using plutonium as fissionable material. The bomb was 60 inches wide and 128 inches long (1.5 x 3 meters) and had a yield of 20,000 tons of high explosive—about the same as the gadget and the Hiroshima bomb.

205. Atomic energy used peacefully.
This 1955 postage stamp commemorates a program proposed by U.S. President Dwight D. Eisenhower in 1953 for reducing the threat of nuclear war through international cooperation in developing civilian uses of atomic energy. The American initiative produced cooperative agreements between various countries.

203. Underwater A-bomb test. To determine the effects of an atom bomb blast on naval vessels, the United States carried out two tests with surplus ships in mid-1946 at Bikini Atoll in the Pacific Ocean. In the second test, shown here, the explosion lifted up a million tons of water, along with entire ships, into the mushroom cloud, whose stem was a mile high. In psychological self-defense, an awed world named a scanty new swimsuit after the event.

204. First hydrogen bomb test. Hundreds of times more powerful than the atomic, or fission, bomb is the thermonuclear bomb, based on the fusion of hydrogen nuclei to form helium. This photograph (taken from a much greater distance than the one above it) shows the first actual explosion of a thermonuclear device, at Eniwetok Atoll in the Pacific on November 1, 1952. The yield was estimated at several million tons of TNT; the blast obliterated a small island and produced a crater more than a mile wide. Within a year the Soviet Union staged its first thermonuclear test.

206. Oak Ridge research reactor. This 30-megawatt reactor at the Oak Ridge National Laboratory in Tennessee was used until 1987 for isotope production, materials testing, and basic physical research.

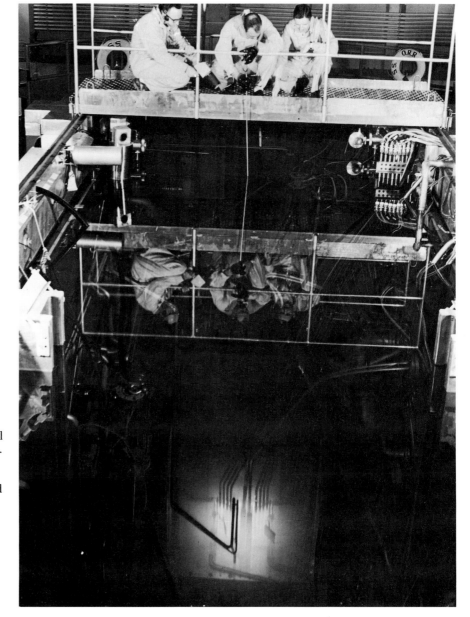

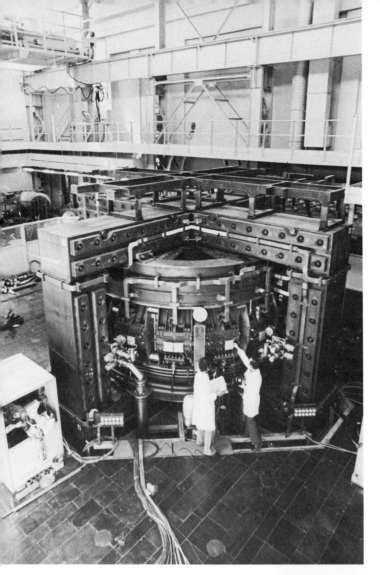

207. Harnessing fusion: the Soviet Tokamak-10. Besides developing uranium reactors for nuclear power generation, researchers sought to harness fusion as a source of clean power. One path toward controlled nuclear fusion uses a doughnut-shaped magnetic "bottle" to generate and confine atoms in the exceedingly high temperatures required for ignition. Such systems are called tokamaks, from the Russian for "toroidal chamber with an axial magnetic field." First proposed in 1950 by the Soviet physicists Igor Tamm and Andrei Sakharov, they were intensively studied at the Kurchatov Institute of Atomic Energy in Moscow. The experimental Tokamak-10 went into operation at the Kurchatov Institute in 1975.

208. Project Sherwood. The first U.S. program in controlled nuclear fusion was Project Sherwood, established at Los Alamos in 1952. By 1957 it generated its first laboratory thermonuclear "plasma," with a temperature of about 10,000,000° Celsius. Shown in the photograph is the toroid and surrounding magnets used in Project Sherwood to confine the plasma.

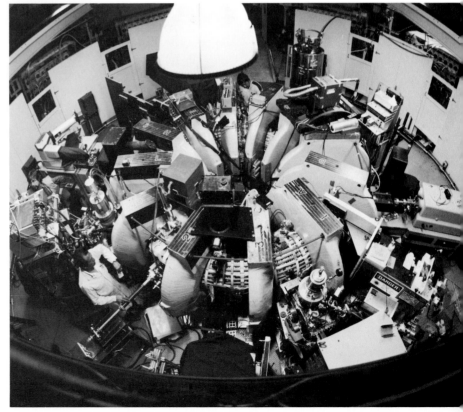

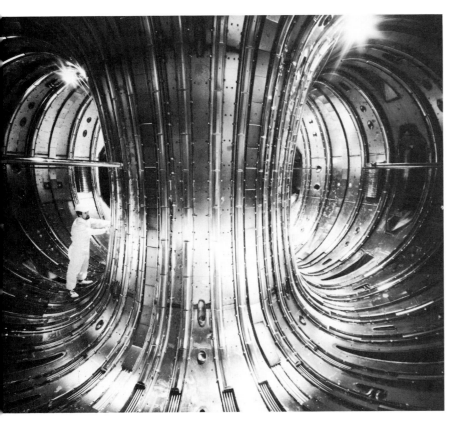

209. JET tokamak: inside the vacuum vessel. The experimental fusion machine at Britain's Culham Laboratory went into operation in 1983. This large vacuum vessel in the center of the machine held the high-speed, highly ionized atoms necessary for fusion. Temperatures of 20,000,000° Celsius were reached in the tokamak's first year. The so-called Joint European Torus was a collaborative project of several European countries.

210. Tokamak Fusion Test Reactor. Another of the experimental machines built in the 1980s was this one, at the Princeton Plasma Physics Laboratory in Princeton, New Jersey. Ionized deuterium (heavy hydrogen) is injected into this doughnut-shaped vacuum chamber and held within the toroid by powerful magnetic fields. In 1982 the Princeton tokamak set a world record for the longest energy-confinement time ever obtained in such a toroidal configuration.

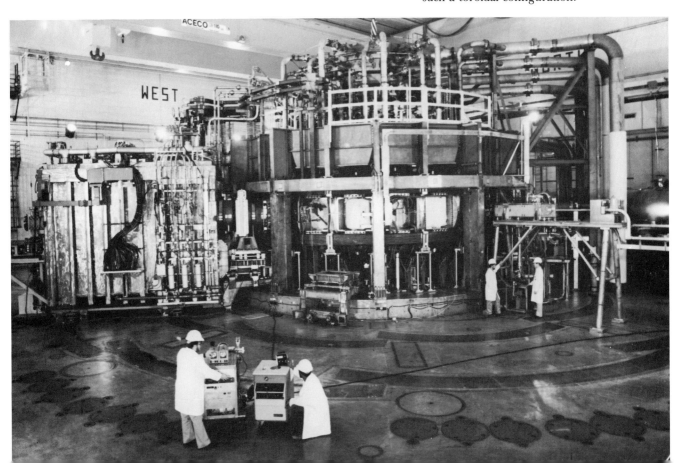

211. Cornell electron synchrotron. The size of the ring tunnels in modern accelerators—nowadays measured in kilometers or miles—makes special transportation a high priority. In this 1968 photograph Cornell University physicists Hans Bethe and Boyce McDaniel use bicycles to make their inspection tour through the magnet ring of Cornell's new "alternating-gradient" synchrotron, which, with a circumference of 630 meters (over 2,000 feet), could accelerate electrons to 12 gigaelectron volts (12 billion electron volts). In the late 1970s an electron storage ring was added in the same tunnel, making possible the study of electron-positron collisions.

14

Inside the Nucleus

The exploration of the atomic nucleus began in 1917 when Ernest Rutherford found that the collision of alpha particles with nitrogen atoms caused the nitrogen to be transmuted into oxygen. Rutherford called for artificial sources of high-energy particles with which to bombard nuclei. His appeal was followed about a decade or so later by the development of several particle accelerators. These included, in addition to Ernest Lawrence's cyclotron (see Chapter 12), the electrostatic generator, invented by the American physicist Robert Van de Graaff, and the linear accelerator, invented by the Swiss physicist Rolf Wideröe and perfected by Lawrence and his colleagues at the University of California at Berkeley. In Rutherford's Cavendish Laboratory at Cambridge, England, John Cockcroft and E. T. S. Walton built a voltage multiplier that in 1932 became the first machine to transmute one atom into another.

Modern big-machine physics arose from the prosperity enjoyed after World War II by nuclear physics, which had produced the winning weapon in the atomic bomb. New inventions included the synchrotron, conceived by Edwin McMillan in the United States and Vladimir I. Veksler in the Soviet Union during the last year of the war. McMillan built an electron synchrotron at Berkeley in 1949. In the early 1950s proton synchrotrons were constructed at Berkeley and the Brookhaven National Laboratory (Long Island, New York) in the United States and at Birmingham in England.

The same period saw the solution of the out-

standing problems in quantum electrodynamics—that is, the quantum theory of electromagnetic phenomena, such as the interaction of charged particles with light or with one another. These advances prompted new efforts by theorists to deal with the strong and weak interactions observed only in nuclear processes.

As accelerators proliferated, so did detectors for trapping the products of the nuclear reactions they induced. The result was an explosion in the number of known particles in the 1950s and 1960s. The first particle to be produced artificially was the pion, which Lawrence's 184-inch cyclotron manufactured in 1947. Another major step forward was the detection of the neutrino. In the early 1930s Wolfgang Pauli and Enrico Fermi had postulated the existence of the particle in order to account for the missing energy in the process of beta decay. For years most physicists considered the neutrino undetectable. Nuclear explosions and reactors, however, were thought to yield massive amounts of the particles. This led Frederick Reines and Clyde Cowan of the Los Alamos Scientific Laboratory to invent a detector in 1952. They used it, or refinements of it, to find the neutrino in the vicinity of large reactors developed to produce plutonium for nuclear weapons.

These early successes stimulated the search for other particles predicted by theory but not yet found. The Berkeley Bevatron produced the antiproton in 1954; the particle was identified with a unique detector system—a combination of coincidence and Cherenkov detectors designed to screen antiprotons from the tens of thousands of pions that accompanied them—that had been invented by Emilio Segrè and his associates. Confirmation, with more traditional emulsion detection techniques, came in the same year.

New detectors, invented to deal with the massive fluxes of particles produced by the giant accelerators, quickened the rate of discovery. The liquid-hydrogen bubble chamber, developed by Luis Alvarez and his collaborators at Berkeley, made possible the identification of hundreds of new subatomic species. The discoveries contributed to new theories of nuclear composition. The fundamental particles of the old nuclear physics were now seen to be composed of "quarks" with properties like "strangeness" and "charm." After the discovery of the J, or psi, particle at Brookhaven and Stanford in 1974, the quark model was established.

The parallel growth of accelerators and detectors dictated the establishment of new laboratories devoted to the art of particle physics. The first was CERN, founded by a group of Western European nations near Geneva in the early 1950s. (The acronym comes from the French form of the organization's original name: European Council for Nuclear Research.) The new laboratories had accelerators that could impart to protons energies of tens of billions of electron volts or more, and they had innovative detectors like spark chambers and heavy-liquid bubble chambers. With this equipment they discovered more particles predicted by theory. For example, CERN in 1983 found the W intermediate-vector boson, which had been postulated in so-called gauge theories of the weak interaction. As CERN expanded, more specialized accelerator laboratories proliferated in Europe. CERN, however, retained pride of place in Western European high-energy particle physics.

The first national laboratory in the United States devoted exclusively to high-energy physics was the Stanford Linear Accelerator Center (SLAC), where a half-mile-long linear electron accelerator was completed in 1966. The second was the Fermi National Accelerator Laboratory, or Fermilab, at Batavia, Illinois. The location of the lab was the subject of intense competition among potential sites in the late 1960s. Completed in 1972, Fermilab's proton synchrotron was by 1985 upgraded through the use of superconducting magnets so that it could accelerate protons to energies of more than 1 trillion electron volts.

As accelerators grew, their design changed. In the early fixed-target configurations, particles were directed into materials held stationary. In the later colliding-beam machines, beams of accelerated particles were directed against each other. Meanwhile, researchers were using a greater variety of particles. Interest in interactions between the lighter atomic constituents, or leptons, led to the construction of electron-positron colliders to supplement the proton-antiproton machines. In a similar way, more sophisticated detectors, like streamer chambers and the exotic time projection chamber, supplemented spark and bubble chambers and scintillation counters.

New questions were raised by the development of the so-called grand unified theories, which combined the theories of the electromagnetic, weak, and strong interactions. Was, for example, the lifetime of the proton truly infinite? It had been thought to be so but was now believed to be finite, albeit enormously great. The evolution of neutrino detectors into ever larger structures provided the means to tackle the problem. In the late 1970s physicists launched searches for evidence of proton decay in tanks containing enough protons to reveal decays that might occur only once in 10^{34} years.

Since the 1920s, experimental and theoretical physicists probing the nucleus have continually recast the notion of "fundamental particles." With increasingly large machines and increasingly powerful theoretical attacks on the citadel of the nucleus, they have sought to push back the frontiers of knowledge to ever smaller, more fundamental entities.

Robert W. Seidel

212. Observing the neutrino: first detector. Predicted in the 1930s, the elusive neutrino, with zero charge and little or no mass, was finally detected in the 1950s by the American physicists Frederick Reines and Clyde Cowan. Their first, tentative observation was made in 1953 with this liquid scintillation detector.

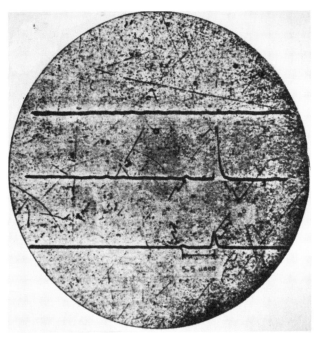

213. Observing the neutrino: the trace. Using a more sophisticated detector, Reines and Cowan made a definitive observation of the neutrino in 1956. Each oscilloscope track here represents a detector tank; the pulses permit identification of the particles involved (a positron and a neutron), from which the intermediate presence of a neutrino was deduced.

214. Detection of the antiproton. The positron, the antiparticle of the electron, was detected in 1932, but not until over two decades later was another antiparticle found experimentally. In 1955 physicists using the Bevatron accelerator at the University of California at Berkeley finally succeeded in producing antiprotons. Their first record of the annihilation of a proton by an antiproton (entering from above on the track marked L) was this "star" in a photographic emulsion.

215. Theoreticians confer. Leading theoretical physicists met in 1947 at Shelter Island, New York, to exchange ideas that helped to identify and solve the developing problems of particle physics. From left to right are Willis Lamb, Abraham Pais, John Wheeler, Richard Feynman, Herman Feshbach, and Julian Schwinger. Robert Oppenheimer called the meeting "the first serious and intimate conference after the war."

216. Cosmotron. This, with its start-up in 1952, was the first accelerator to raise particles to energies of more than a billion electron volts. It was also the first proton synchrotron; its magnetic field was synchronously increased as the protons were accelerated, so that the particles' orbits stayed more or less the same. Using only an outer ring of magnets, the Cosmotron was much more economical than a cyclotron. The machine, at Brookhaven National Laboratory on Long Island, New York, was taken out of operation at the end of 1966.

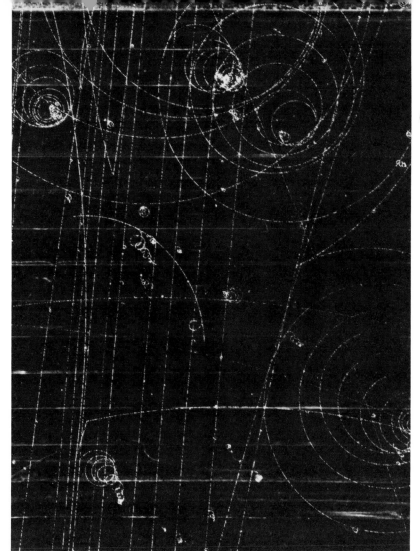

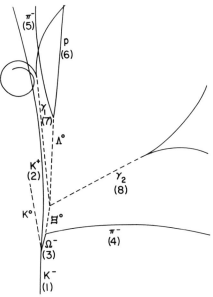

217. The omega minus. Of the many new particles found at mid-century, the most famous was the omega minus predicted by the so-called "eightfold way" theory, which classified particles into groups according to their symmetrical properties. At left is the first bubble chamber photograph showing the particle's existence, taken at the Brookhaven National Laboratory in 1964. The diagram above identifies the particle tracks. The omega minus (3) was produced when a K-minus meson (1) collided with a proton.

218. The discovery apparatus. The "80-inch" liquid-hydrogen bubble chamber in which the omega-minus particle was found is surrounded by the copper coils and iron yoke of a 400-ton electromagnet. Particle collisions in the chamber were photographed by cameras through openings above the staircase at the left. The apparatus took 250 man-years to design and build, at a cost of about $6 million.

219. Linear electron accelerator.
The vacuum tube in the Stanford Linear Accelerator Center is 3.2 kilometers (2 miles) long and is aligned to half-millimeter accuracy. Electrons, fired from a "gun" at the far end of the straightaway, are driven by radio waves in a split second to the experimental areas in the foreground of the picture. Around 1970, SLAC's electrons provided the first intriguing glimpses of the world within protons and neutrons. Thus, within its first years of operation, the California laboratory became an international center for particle physics.

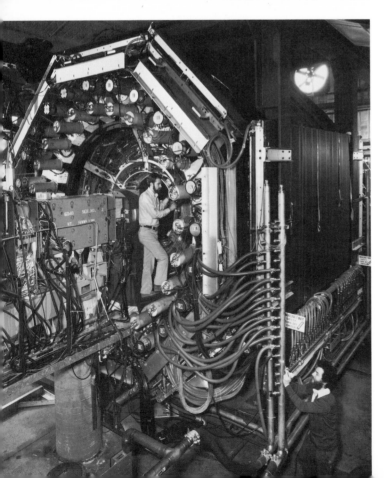

220. Multipurpose detector for a particle collider.
Barely had the linear accelerator been completed at SLAC when physicists won permission to add an electron-positron collider ring. In order to study the annihilation of matter by antimatter, they built this general-purpose particle detector, the Mark I. Sixteen layers of wire spark chambers—100,000 wires in all—are packed within an electromagnetic coil, and the trajectories of the collision fragments are reconstructed by computer. With the Mark I, SLAC's scientists discovered the psi particle, the tau lepton, and charmed particles.

221. Intimations of charm. Here, grouped together in one corner of their apparatus at the Brookhaven National Laboratory, stands the team headed by Samuel Ting (*far right*) that in 1974 discovered a new particle in proton-beryllium collisions. Ting called it the J particle. Almost simultaneously, a group at Stanford headed by Burton Richter found the same particle in electron-positron annihilations. They named it the psi particle. Subsequently known as the J/psi, the heavy particle had a longer life than expected for its particular mass-energy. Because of this, theorists suggested that the particle was built of a new kind of quark, a "charmed" quark.

222. Proof of charm. In 1964 the American physicist Murray Gell-Mann had proposed that neutrons or protons consist of simpler units, which he called quarks. The J/psi was explained in terms of a charmed quark and its antiquark. Experimentalists soon found evidence for particles consisting of a charmed quark and other quarks. An actual track of such a charmed particle was not detected until 1977, on this stack of photographic emulsions from the Big European Bubble Chamber at the CERN laboratory near Geneva. Here the long straight track is from a charmed, positive lambda-zero particle. It begins with a neutrino interaction at the left star and decays into a proton and two other particles at the right. The track is actually only 0.35 millimeter long, and the decay time was nearly a million millionth of a second.

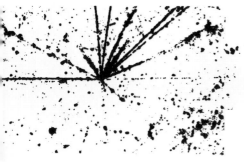

223. CERN. In this mid-1980s aerial view of the CERN laboratory, the large circle marks the tunnel for LEP, a large electron-proton colliding beam machine that was being constructed under the French-Swiss border. The ring is 27 kilometers (about 17 miles) in circumference. The smaller circle marks the underground position of CERN's Super Proton Synchrotron.

224. UA1 detector. Along with bigger and bigger accelerators, capable of accelerating particles to high energies, there came larger and more sophisticated detectors. The $30 million colossus shown here is two stories high and weighs as much as five jumbo jets. Located in "Underground Area 1" of the Super Proton Synchrotron at CERN, it recorded the shattered nuclear fragments from the collision of protons with antiprotons; in particular, it was used in the 1983 detection of the W and Z particles.

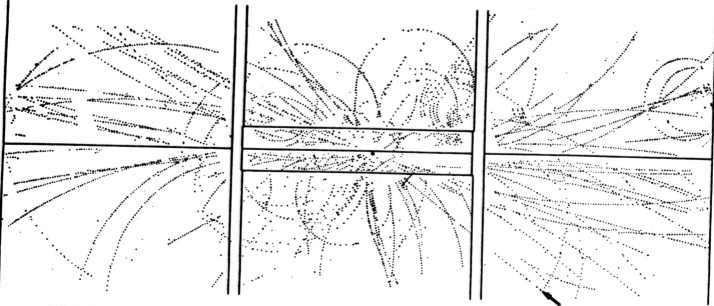

225. W boson. Predicted by the theory of the so-called weak nuclear force, the W and Z particles became the goal of a frenzied experimental search. In this computer output from the CERN UA1 detector, the track indicated by the arrow is a high-momentum electron produced in the decay of the W particle.

226. Experimental hall for heavy nuclei. Guided by magnets, a beam of accelerated particles is distributed to more than a score of different experimental devices in the GSI facility near Darmstadt in West Germany.

227. Serpukhov synchrotron.
The world's most powerful atom smasher for five years after it began operating in 1967, this huge proton accelerator fueled international competition for ever larger machines. Located near Serpukhov, south of Moscow, it is an alternating-gradient machine about 1,500 meters (nearly 1 mile) in circumference and was designed to accelerate protons to a maximum energy of 76 gigaelectron volts. Unfortunately, it never succeeded in generating a major breakthrough.

228. Fermilab. The American response to the Soviet competition was the proton synchrotron at the Fermi National Accelerator Laboratory in Illinois, which achieved a record-breaking 200 gigaelectron volts in 1972. Its main ring, 6.3 kilometers (3.9 miles) in circumference, is visible from space. Converted to a proton-antiproton collider, it reached a record 1,600 gigaelectron volts in 1985.

229. The search for proton decay. In the late 1970s physicists began exploring the possibility that the proton might be subject to decay. Theorists suggested an almost inconceivably long half-life, but in a sufficiently large mass a few protons could self-destruct each year. Huge tanks of water provided researchers with a compact source of protons. Here, in a salt mine near Cleveland (underground to shield against cosmic rays), a diver installs detectors. Unexpectedly, the array found neutrinos given off in the core collapse of Supernova 1987A in the Large Magellanic Cloud.

Part Five

THE
STRUCTURE
OF MATTER

230. Nobel's laboratory. Alfred Nobel, the nineteenth-century Swedish industrialist who established the Nobel prizes, was himself a chemist and engineer. The inventor of dynamite and other explosives, he held more than 350 patents. In the last years of his life, Nobel moved to a mansion near San Remo, Italy. On the grounds surrounding the villa he built this laboratory for his work, which included the development of "progressive smokeless powder." His researches also involved electrochemistry, optics, and physiology. A multimillionaire, Nobel left the bulk of his fortune to a foundation for awarding prizes "to those who, during the preceding year, shall have conferred the greatest benefit on mankind." The awards in physics, chemistry, and medicine or physiology have provided a roster of the greatest discoveries in those disciplines, and no prize carries greater prestige in the physical sciences.

15

*H*ow *A*toms *U*nite

Chemists long investigated the gross phenomena associated with the formation of molecules without knowing how atoms unite. Their situation was like that of Charles Darwin, who articulated his theory of evolution without any knowledge of genes or the mechanism of inheritance. The study of organic, or carbon-based, compounds in the nineteenth century, however, put a premium on the search for molecular structure, since many of these substances were isomers—that is, they had exactly the same atomic constituents but behaved differently because they differed in structure. Thus, when August Kekulé declared in 1865 that the six carbon atoms in benzene formed a ring, chemists began to attend not just to the numbers and types of atoms present in a compound but also to the atoms' arrangement. Following Louis Pasteur's pioneering work on the ability of certain organic acids to rotate polarized light, Jacobus Henricus van't Hoff in the Netherlands, J. Achille Le Bel in France, and Johannes Wislicenus in Germany founded stereochemistry—a specialty dealing with atomic arrangement—by probing the asymmetry of the carbon atom and the peculiarities of its double bond. By 1900 the practice of making models of molecules out of balls and sticks had become commonplace.

Practical rewards accompanied the intellectual challenges involved. Alfred Nobel, for example, used chemical expertise to perfect techniques of making nitroglycerin; he then bound the explosive liquid with other materials to yield his infamous dynamite. In organic chemistry especially, such innovative entrepreneurs made fortunes via chemical manipulations. Sugar refiners and German dye firms led the way; the latter mushroomed around the turn of the century from small family affairs to giants with thousands of workers. Industrialists rewarded professors handsomely for consulting work, and students, lured by the commercial successes, eagerly sought chemical credentials. Nobel left most of his fortune to endow the prizes that bear his name. Unparalleled prestige was conferred on the likes of van't Hoff, who took the first chemistry prize, awarded in 1901, and on Germany's Emil Fischer, the second recipient of the chemistry prize. Fischer was renowned for elucidating the complex structure of sugars, for synthesizing many organic compounds, and for turning out dozens of productive research biochemists.

Subsequent achievements followed opposite paths: one toward understanding atomic-level interactions; the other, huge molecules. J. J. Thomson's discovery of the electron in 1897 prompted chemists and physicists alike to investigate not just how the various atomic nuclei in a molecule are arranged relative to each other but also what interconnects them and how such bonds come to be made. This proved especially enticing to physical chemists; they were concerned about the attraction and repulsion of atoms at least since the 1880s, when the Swede Svante Arrhenius proposed that a salt breaks up into charged bits of molecules called ions when it dissolves in water.

Thus, by the close of the nineteenth century chemists had begun to picture atoms as something more than the hard, spherical balls postulated decades before by the English man of science John Dalton and had become convinced that electrons figured somehow in molecule formation. Only five years after Thomson's electron announcement, Gilbert Newton Lewis in the United States pioneered a heuristically valuable cubic model of the atom with an outer shell of up to eight electrons that could interact with electrons on the outsides of other atoms. Subsequent refinements included the ionic bond, in which one atom bonds to another by giving up one or more electrons to it, and the covalent bond, in which two atoms complete their outer shells by sharing electrons.

The rudimentary cubic schemes for the outer electron shells eventually fell victim to the revelations of quantum mechanics, but modern ideas of an electrically active outer shell date from this period. By the 1920s sophisticated chemists were thinking of the shells more in terms of "clouds" of whirling electrons. The American theorist Linus Pauling drew attention to the tendency of such electrons in a bond to favor one of the atoms around which they resonate, thereby creating mildly positive and negative zones on the molecule. Especially pronounced in the tiny hydrogen atom, this phenomenon allowed exposed hydrogens in certain molecules to form fairly weak "hydrogen bonds" between themselves and "electronegative" zones on nearby atoms. Sometimes developing between molecules, these links helped explain the peculiar properties of water and would later figure centrally in models of the structure of DNA and other complex organic molecules.

The early attention to small molecules that could be visualized as clumps of cubes, however, prejudiced chemists against the idea that extremely large molecules could exist. The study of colloids—gluelike substances that could not be strained through fine filter paper—encouraged the notion that such large lumps of matter were multimolecular aggregates called micelles, not single molecules with thousands of atoms. Battling considerable skepticism, the German chemist Hermann Staudinger steadily gathered evidence for the alternative view that substances like natural rubber were actually huge macromolecules, or polymers, assembled by linking smaller units together into a chain. Although his conception of polymers as rigid and rodlike was supplanted by the flexible models of the Austrian Herman Mark and the Swiss Werner Kuhn, Staudinger's macromolecular hypothesis survived. Indeed, it was crucial to the intensive work on synthetic rubber during World War II and to the spectacularly successful postwar plastics industry. It also paved the way to understanding biological macromolecules. Armed with the new insights, and aided by armies of assistants, astute chemists like the American Robert B. Woodward began around 1940 to synthesize unprecedentedly complex organic compounds, including quinine, chlorophyll, cholesterol, and vitamin B_{12}.

A key to such successes was the increasing sophistication of electrical and electronic apparatus for the analysis of chemical substances. Early equipment merely measured gross physical properties, such as conductivity. More sophisticated devices, developed later, used the physical behavior of molecules as an aid in determining their composition and structure. Infrared spectrophotometers, for example, could identify the bonds in a compound by measuring the frequencies of infrared light absorbed by its molecules and translated into vibrations along their bonds. Molecular beam devices were similar; they recorded the absorption of weak electromagnetic radiation by a stream of molecules moving through a strong magnetic field and used the pattern of absorption to determine molecular structure. This technique, called nuclear magnetic resonance, was largely the work of the American physicist I. I. Rabi. Physicists began to collaborate more and more with chemists as quantum electronic principles came to loom large in chemistry. In time, other devices were developed, such as unusual forms of electronic microscopes that could make images of individual atoms, achieving resolutions of less than 1 angstrom (about one five-millionth the width of a human hair).

With increasingly sophisticated understanding of how atoms unite, chemists refined to an art their skill at constructing molecules. They became capable, for example, of making intricate cubic molecules that entrap benzene rings inside themselves. The 1980s saw the development of peculiar ceramic substances that became superconductors at much warmer temperatures than accepted theory permitted. Such continuing breakthroughs show that there is still much to learn about the links between atoms.

P. Thomas Carroll

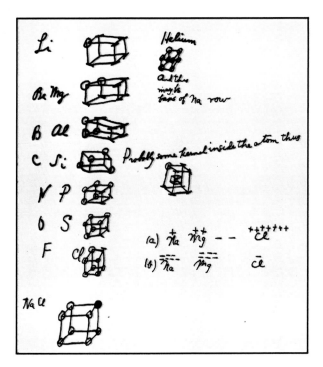

231. The cubic atom. Early in the century the American chemist Gilbert Newton Lewis visualized the atom as a series of concentric cubes, with an electron possible at each of the eight corners. This "octet" model was in accord with the repeating eight-element cycle in the periodic table. The diagrams shown here are from a memorandum he prepared in 1902 outlining his theory. The sketches at far left show only the outer electrons—the ones that, as he would propose, participate in chemical bonding. Note that this theory came before Rutherford's discovery of the positive nuclei of atoms.

232. Pauling's hydrogen bond. These two views of ice show the so-called hydrogen bond—the weak attractive force linking together the molecules of water. At left the larger spheres depict oxygen atoms, each tightly bound (at an angle) to a pair of hydrogen atoms. The exposed flank of the oxygen is slightly negative and forms a weak bond with the slightly positive side of a hydrogen atom. Repeated symmetrically, these hydrogen bonds form a solid ice crystal. The American chemist Linus Pauling showed how such bonds play a major role in the structures of, for example, proteins. The right-hand side shows the molecules in roughly their correct relative size.

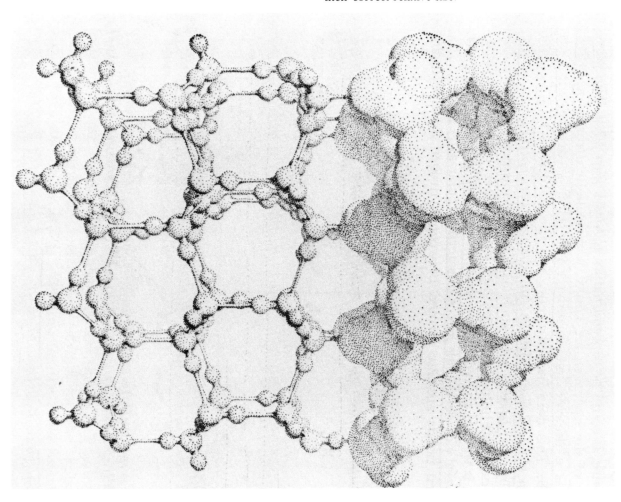

233. Fischer. The German chemist Emil Fischer won the Nobel Prize in chemistry in 1902 for his research on sugars and purine substances. He realized that uric acid and related compounds that he synthesized (including caffeine) were oxides of a ring molecule that he called purine. He synthesized well over a hundred derivatives, including the nucleotides adenine and guanine, which are now known as rungs in the DNA ladder. Fischer's work on organic chemistry laid the foundations for the development of biochemistry. In addition, he exerted an enormous influence by training the younger generation of organic chemists.

235. Mark with Staudinger's macromolecule. Staudinger thought that polymer chains were long and rigid, but in the late 1920s Herman Mark and others favored a more flexible molecule—a view that proved to describe most polymers accurately. In this 1986 photograph Mark holds a model of his old opponent's rigid macromolecule.

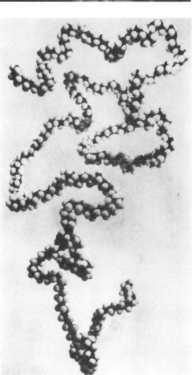

236. Kuhn's coils. The Swiss chemical physicist Werner Kuhn saw the molecule as a chain of rigid links with flexible joints; in solution the polymer was a compact coil—a view closer to the truth than Staudinger's idea of rigid chains. Shown at left is Kuhn's 1934 conception of a macromolecule.

237. Woodward. Holding a model of a synthetic protein analog, in a late-1940s photograph, is American chemist Robert Woodward. One of the foremost organic chemists of his generation, he synthesized the antimalarial drug quinine (1944), cholesterol and cortisone (1951), strychnine (1954), the tranquilizing drug reserpine (1956), chlorophyll (1960), and vitamin B_{12} (1972).

234. The macromolecular debate begins. Early in the century polymers were generally believed to be aggregates of small molecules linked together by weak intermolecular forces. In this 1920 paper, "On Polymerization," the German chemist Hermann Staudinger proposed they were giant molecular chains. Not until 1953 was his work recognized with a Nobel Prize.

238. Analytical equipment. This piece of apparatus, from a 1914 catalog, was used to determine the electrical conductivity of electrolytes. It is hardly a coincidence that electrification of instruments entered the laboratory at the same time as physical chemistry and the same time that interest in ionization emerged as a crucially important specialty.

239. Molecular beam apparatus. Molecular beams—streams of molecules moving in the same direction—have proved a fruitful tool for chemists and physicists since the first such experiment, by the French scientist Louis Dunoyer, in 1911. In the second half of the century colliding molecular beams provided data on reactions between molecules. The apparatus below, photographed at a Harvard physics laboratory in 1950, was used to investigate nuclear magnetic resonance.

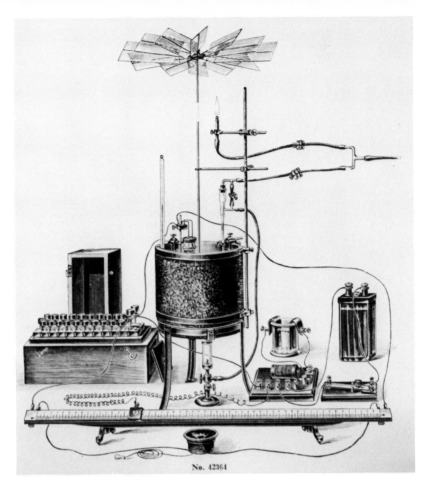

No. 42364

240. Hot lab. Scientists working with highly radioactive substances developed precautions to safeguard against radiation poisoning. Typically an object or substance under study was kept behind a lead wall and was manipulated by remote control. In the apparatus shown here, photographed at mid-century, the work being done could be observed in the mirror at upper left. Some stages of the isolation of element 96 (curium), the third transuranium element to be discovered, were carried out in this laboratory during World War II.

241. Discovery of lawrencium. Like the equipment above, these remote-controlled instruments were used to handle radioactive materials. Placed at the end of a heavy-ion linear particle accelerator at the University of California at Berkeley, the apparatus was used in 1961 in the discovery and identification of transuranium element 103, named lawrencium after Ernest Orlando Lawrence, the pioneer builder of cyclotrons.

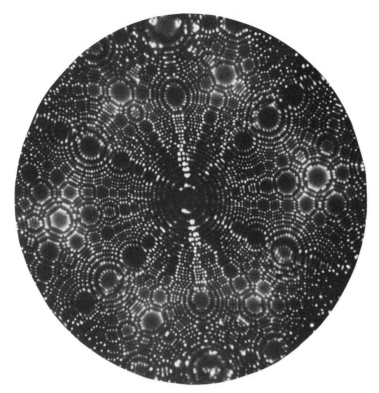

244. A single atom at rest. How to pin down, and hold at rest for observation, a single member of a cloud of chaotically moving atoms—this tantalizing problem was solved for the first time in 1979 by a team of physicists in Heidelberg. The arrow points to a single barium ion that was immobilized using laser cooling. The kinetic temperature, a measure of the motion of the ion, was a tiny fraction of a degree above absolute zero.

242. Imaging atoms. The type of electron microscope known as the field-emission microscope, developed by the German-born physicist Erwin W. Müller in the 1930s, had by 1955 been sufficiently refined to yield the first images of atoms. Heavier metals, such as the tungsten shown here, were the first elements to be imaged. What we see is a nearly hemispherical crystal with a radius of about 560 angstroms. (One angstrom is one ten-billionth of a meter.) Each tiny spot is one atom of tungsten.

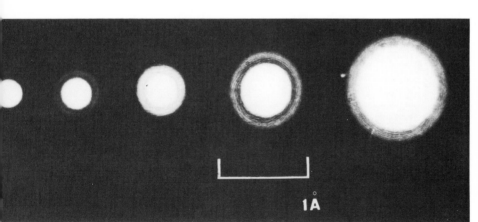

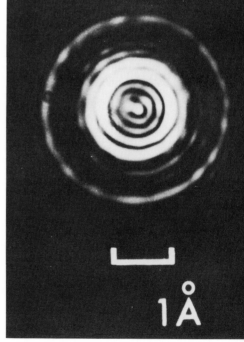

243. Imaging the atom: electron-wave holography. These pictures, taken in the mid-1970s with a holographic microscope, constitute a milestone in magnification and resolving power. Above are five pictures of argon, the first atom to be so imaged. At right is arsenic pentafluoride, the first molecule imaged with the technique. The outer ring represents the shell of five fluorine atoms around the central arsenic atom. The image seen actually depicts the average of an ensemble of identical particles.

245. Superconductivity: the Meissner effect. The phenomenon of superconductivity—the loss of electrical resistance at low temperatures—offers intriguing possibilities for carrying electric currents efficiently. Among the curious ancillary properties of some superconducting materials is an effect discovered in the 1930s by the German physicists Walther Meissner and R. Ochsenfeld. When a material is cooled so much that it becomes a superconductor, it not only offers zero resistance to the flow of electric current but also expels any magnetic field that is present. The superconductor becomes a "diamagnet," that is, a substance that is repelled by either pole of a magnet. At right, a lead sphere, normally a rather poor conductor, is not only superconducting in the cold bath of liquid helium (at a temperature of 4.2 degrees absolute) but is freely supported by its repulsion to opposing magnetic fields. Beginning in 1986 physicists discovered classes of ceramic compounds that become superconducting at much higher temperatures—a finding that opened up greater prospects for the application of superconductivity.

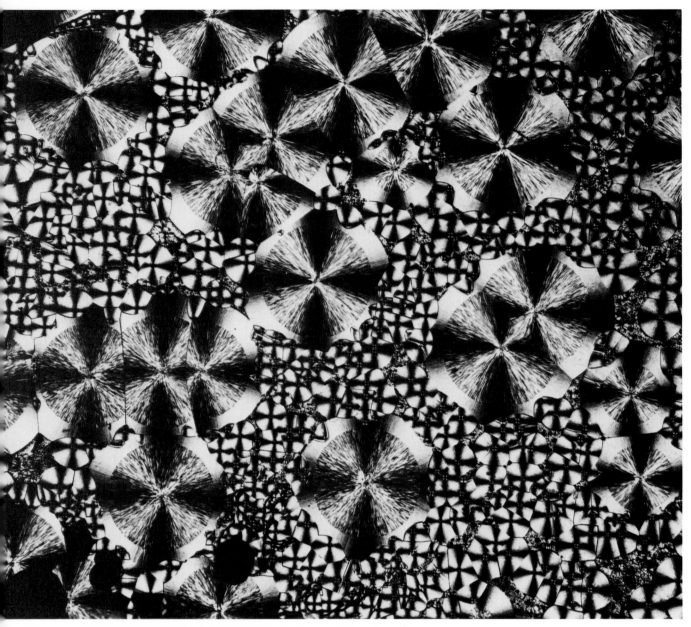

246. Polymer crystal structure. Products made of giant polymers—
rubbers, fibers, films, glues, plastics—display a bewildering variety of physical properties. Some are completely amorphous, while others, such as this
experimental siliconelike polymer, made in the General Electric Research
Laboratory, have a partly crystalline structure, which produces rigidity
without brittleness. Under polarized light the polymer reveals the intricate
pattern we see in the photomicrograph above. Each center is a cluster of
minute crystals, or crystallites. Within and between the centers, the polymer's structure is amorphous, or uncrystallized. At mid-century, when this
silicone analog of Dacron was under study, chemists claimed to understand
the viscoelastic behavior of polymers more clearly than that of metals.

16

*C*hemical *T*echnology

In ordinary life a vast distance separates the world of clothing and fashion from that of war and weapons, but for the chemist the distance is very small. This lesson was driven home quickly and dramatically in 1914 when the Allied powers found themselves at war not only with the German Empire but also with the German Dye Trust. A handful of chemical firms, including BASF, Bayer, and Hochst, were able in the first years of the century to monopolize the manufacture of synthetic dyestuffs and the coal-tar chemicals that went into them, such as toluol, benzol, and naphthalene. These "intermediates" were necessary for the manufacture of such militarily indispensable materials as smokeless powder, picric acid, and trinitrotoluol (TNT). The cutoff of German organic chemicals at the beginning of World War I forced other European nations and the United States to create their own synthetic chemical industries, thereby laying the foundations for modern industrial chemistry.

The source of German chemical power in the late 1800s had been the combination of well-supported chemical education and an industry ready and willing to invest time and money in wide-ranging research. Indeed, German firms provided the model for the industrial research-and-development laboratories that were to transform the relationship between science and business in the new century. Similarly, Heinrich Caro, BASF's director of research from 1868 to 1889, typified the modern research director. His dogged pursuit of a synthetic

indigo, stretching over almost two decades, was an object lesson to the modern industrialist who wanted to stay ahead of his competitors. The products of similar efforts were transforming almost every sphere of life. From drugs like Bayer's aspirin and Paul Ehrlich's Salvarsan to the ionone and terpineol that became indispensable to the twentieth-century perfumer, the fruits of German chemical laboratories made the last decades of the nineteenth century and the first one of the twentieth a heroic age of applied chemistry.

The chemical heroics were not restricted to so-called "fine" chemicals. Large-scale chemical manufacture also changed in the new century. The Haber-Bosch process for synthesizing ammonia from atmospheric nitrogen broke a chemical bottleneck, transforming the arts of war (by providing a major constituent of explosives) and of peace (by affording relief from dependence on natural sources for nitrogen fertilizers). New techniques were also devised for large-scale production of gasoline and other products from crude petroleum. In the United States, William Burton introduced his thermal "cracking" process just before World War I. At about the same time, the first catalytic cracking techniques were invented, but these were not practical until the work of Eugène Houdry in the 1930s.

To the ordinary man or woman, the most spectacular products of creative industrial chemistry in the first decades of the twentieth century were the new materials, especially the plastics. At the turn of the century most people would have encountered celluloid, the first of the cellulose plastics. Its use in photographic (and cinema) film, in collars and cuffs, and in often elegant but inexpensive toilet goods conveyed some idea of the value of a general-purpose plastic material. Celluloid's flammability, however, was a serious drawback (for both manufacturer and consumer), spurring a search for substitutes. Other cellulose plastics emerged: cellulose xanthate, or viscose, was discovered in 1892 by the English chemists Charles Cross and Edward Bevan and was later developed into rayon. Cellulose acetate was actually first prepared in 1869, but it was not until 1910 that the Swiss brothers Camille and Henri Dreyfus developed successful large-scale manufacture of the nonflammable material. It was rapidly adopted for the emerging movie industry. As war approached, it became a strategic commodity as a doping material for the fabric wings of aircraft.

Another strategic commodity was rubber, and during World War I German chemists tried mightily to devise a substitute that would replace the material so tightly controlled by the Allies. The result was an unsatisfactory and costly polymer of butadiene [$CH_2=CH-CH=CH_2$]. In 1925 the Du Pont Company began an organized attack on the problem of synthetic rubber, reminiscent of BASF's indigo effort. This time, however, it took only six years for the Du Pont chemists (from 1928 under the leadership of Wallace Carothers) to come up with neoprene, a polymer of chloroprene [$CH_2=CCl-CH=CH_2$]. The introduction of neoprene in the midst of the Depression had little economic or political impact, but as the thirties wore on and war clouds gathered over Europe, the search for new synthetic rubbers accelerated. The result was a variety of materials, such as the Buna rubbers, polysulfides (Thiokol), and polyvinyl chloride (PVC), each with useful properties of its own.

By the 1930s, however, the creation of new synthetic materials was no longer unusual. The large, well-funded laboratories of companies like Du Pont were not the only source of such inventions. In fact, the first completely synthetic plastic (that is, not made from natural polymers like cellulose) was created in the backyard laboratory of Leo Baekeland just after the turn of the century. Baekeland, who had received his doctorate from the university in his native Ghent in Belgium, was able to parlay his knowledge into a fortune soon after his 1889 immigration to the United States by selling the formula for Velox photographic paper to George Eastman. Dabbling in interesting chemical problems, Baekeland decided to investigate the reactions between phenol and formaldehyde in hopes of finding a shellac substitute. After discovering how to control the tricky combination, Baekeland came up with a remarkable substance he optimistically (and immodestly) dubbed Bakelite. Bakelite was the first of a long line of synthetic resins whose useful properties were to become essential to the growth of such industries as aviation and electric power.

Baekeland was guided as much by a skilled chemist's hunches as he was by theory in the development of his material. In the following decades, theory and practice continued to move ahead in tandem. New concepts, such as the macromolecular theories of Hermann Staudinger, gained currency at the same time that new materials were emerging from laboratories on both sides of the Atlantic with increasing frequency. In 1930 the chemists of Germany's I. G. Farben put polystyrene on the market, providing an enormously versatile new plastic. A few years later the much smaller German firm of Röhm and Haas introduced its remarkably clear polymer of methylmethacrylate (acrylic) as Plexiglas. The greatest notoriety was attracted by the development at Du Pont of the polyamine nylon under the direction of Wallace Carothers. Nylon turned out to be another versatile material, serving as an "artificial silk" for hosiery or parachutes as well as a molding compound for everything from combs to fishing rods. An additional Du Pont triumph before World War II, though one much slower to gain public attention, was polytetrafluoroethylene (PTFE), which eventually came to market as Teflon. In the years after the war, chemical firms everywhere continued to exploit both new theories and new techniques to produce a flood of artificial materials, transforming the very feel of everyday life.

Robert Friedel

247. Indigo industry. Long an important dyestuff, indigo was obtained only from natural sources until a successful synthetic process was introduced in the late nineteenth century by Germany's Badische Anilin- und Soda-Fabrik. The first commercial synthesis of indigo was carried out at BASF around the turn of the century in the apparatus at right.

248. Indigo laboratory at BASF around 1900. Bringing synthetic indigo to the market took seventeen years of directed research. The project is said to have cost BASF about $5 million, the most that had ever been spent on a single research program. With the help of the indigo project, Germany by 1900 accounted for over 80 percent of the world's production of dyestuffs and was the world leader in chemical technology.

249. Haber's ammonia-synthesis lab equipment. The German physical chemist Fritz Haber devised a method of synthesizing ammonia directly from hydrogen and atmospheric nitrogen under high pressure. The method was developed into a large-scale industrial process by the BASF chemist Carl Bosch. The Haber-Bosch process had huge economic significance: ammonia could be made into nitric acid, used to produce nitrate explosives as well as fertilizer. The development of the Haber-Bosch process was a contributing factor to Germany's launching of World War I. Haber won the Nobel Prize in Chemistry in 1918, and Bosch in 1931.

251. Safe but deadly ribbons. Nitrocellulose, a powerful explosive when impacted, provided most of the firepower in World War I. The key ingredient in smokeless powder, it was more powerful than gunpowder. The demand for smokeless powder for cartridges was enormous, forcing a drastic increase in output. This nitrocellulose production scene is from a Du Pont factory. Du Pont supplied over 40 percent of the standard explosives used by the Allies in the war.

250. First ammonia factory. Under Carl Bosch's direction a BASF ammonia plant was constructed quickly at the village of Oppau along the Rhine; it went into operation in September 1913. The basic process—the catalytic synthesis of ammonia from hydrogen and atmospheric nitrogen—was carried out at top right (below the large gas holders). The two high-chimneyed plants in the middle converted the ammonia into ammonium sulfate. Each of the big concrete silos between the two held 50,000 tons of fertilizer salt. The big research building in the lower left corner was renowned worldwide as the Ammonia Laboratory. Some distance above the research building are buildings (with twenty-one high absorption towers) where the ammonia was converted into nitric acid and saltpeter.

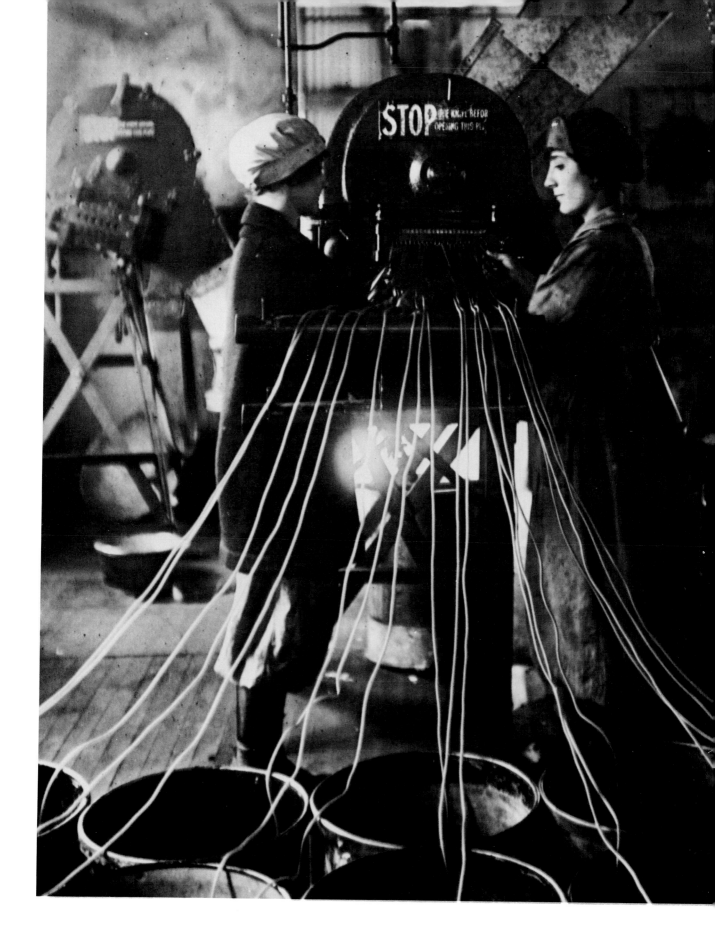

252. Burton cracking stills. The surging popularity of the automobile early in the century meant increasing demand for gasoline. In 1912, Standard Oil researchers headed by the chemist William M. Burton developed a thermal cracking process for producing gasoline from crude oil. Batteries of Burton cracking stills can be seen in this 1914 picture of the Whiting, Indiana, refinery.

253. Carothers. After joining Du Pont in 1928 to head the company's organic chemistry research group, Wallace Hume Carothers not only made significant theoretical contributions but initiated two new industries—synthetic rubber and wholly synthetic fibers, notably nylon. Here he demonstrates neoprene; discovered in 1931, it was the first commercially successful general-purpose synthetic rubber.

254. Continuous cracking stills. Burton's technique was a batch process. Continuous cracking was the next major advance. Prominent in the foreground of this mid-1920s photograph of a Texaco (then the Texas Company) refinery at Port Arthur, Texas, are a large number of Holmes-Manley stills. The first continuous thermal cracking process to be both practical and commercially profitable, the Holmes-Manley process was first put into operation in 1920.

255. Synthetic rubber industry. The United States ▶ became a sizable producer of synthetic rubber in World War II, when the loss of its primary source of rubber—plantations in the Far East—prompted a crash program of research and production to help meet the need for industry and the war effort. Epitomizing the demand for rubber is the heavy-equipment tire being molded in this 1944 picture, taken for *Fortune* magazine by André Kertész.

172

256. Baekeland's notebook. Credit for commercial production of the first completely man-made giant molecule goes to Leo Baekeland, a chemical inventor who emigrated to the United States from Belgium. Following the hints of earlier chemists who abhorred the sticky resinous material that sometimes clogged their equipment, he developed a hard plastic made from phenol and formaldehyde. Baekeland distinguished four stages in the reactions producing the material. Type D, he writes, "is insoluble in all solvents, does not soften. I call it Bakalite." Bakelite's first important use was as an electrical insulator.

257. Baekeland's laboratory. Before Bakelite, Baekeland invented Velox, a photographic paper that could be printed by artificial light. He sold his company and rights to Velox to camera magnate George Eastman for a million dollars and retired to work in this private laboratory, shown here in an 1898 photograph, in the backyard of his Yonkers, New York, home. With the invention of Bakelite, the first thermosetting plastic, early in the twentieth century, he became the founder of the modern plastics industry.

258. Plastics in industry. This World War I photograph reveals another early application of plastics. The fabric of airplane wings was "doped" with a plastic as a waterproof varnish.

259. Plexiglas. The first piece of "organic glass" was cast by German chemist Walter Bauer in 1932. Through the cooperation of German and American technologists, this polymerized methylmethacrylate went into industrial production by Röhm and Haas in Germany in 1935 and in the United States the following year. Strong and lighter than glass, Plexiglas promptly found a place in airplane designs. During World War II it was put to use in cockpit canopies, windows, and windshields. Plexiglas contributed to the trend to streamlining: 360-degree gunner's turrets and bombardier's enclosures appeared on larger planes. In the photograph is a Plexiglas nose cone.

260. Teflon. First marketed in 1944, Teflon, the strong, tough polymer polytetrafluoroethylene, is chemically and physically inert. Famous as a coating for cooking ware—nothing adheres to it—Teflon is also a fine electrical insulator. One of its many applications is as part of the composite of materials in the extravehicular-activities suit worn by U.S. astronauts. Here, in a late 1984 scene, NASA astronaut Dale Gardner works in the cargo bay of the space shuttle *Discovery* in his Teflon space suit.

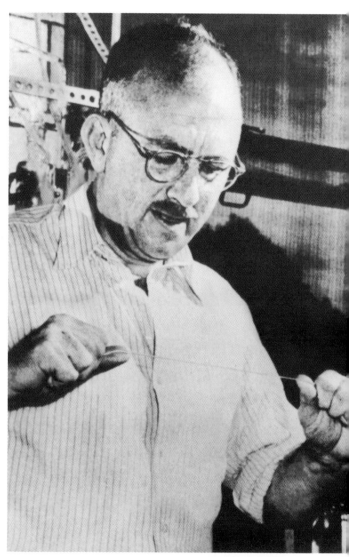

261. Birth of nylon. Chemist Julian Hill, an assistant to Carothers at Du Pont, here reenacts a 1930 experiment creating the first completely synthetic fiber. At left, we see Hill pulling a molten sample of the material from a test tube; the molasseslike mass stuck to the glass stirring rod and was drawn out into a thin fiber.

262. A tough fiber. After the fiber cooled, Hill was able (*above*) to stretch it out to a few times its original length. At a certain point, with the molecular chains stretched out to their ultimate length and lying parallel to one another, the fiber became "fixed," showing remarkable toughness. This experimental material was the forerunner of nylon, but it took $27 million and a decade of work to perfect the process. Nylon was the first synthetic material to be presented not as a substitute for a natural fiber but as a superior product.

263. First Xerox copy. Chester Carlson, the American inventor who developed the electrostatic dry-copying process known as xerography, made this first xerographic print with his assistant Otto Kornei in Astoria, New York, in October 1938. Xerography uses the property of photoconductivity, forming a positively charged electrostatic image on a selenium-coated drum. Negatively charged particles of toner are sprayed onto the drum and then transferred and fused to paper. The process became commercially available in the early 1950s, and the first Xerox copier was introduced in 1960.

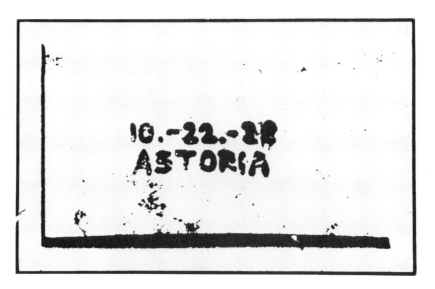

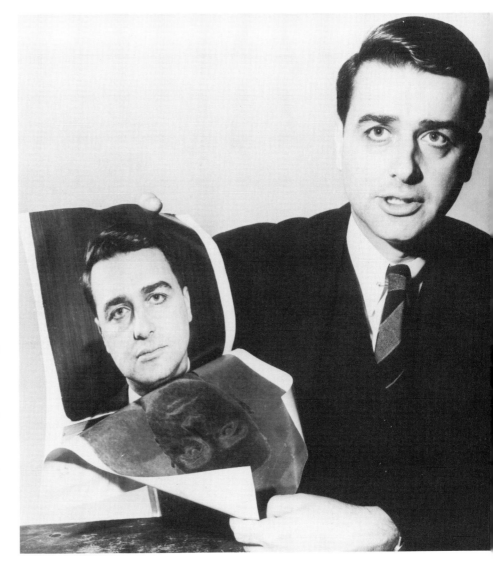

264. Advent of instant photography. Edwin Land, the inventor of the one-step Polaroid Land photographic process, here demonstrates his camera and process in 1947 to the Optical Society of America. He is uncovering the black-and-white photograph of himself just one minute after it was taken. The process produced the greatest revolution in photography since the introduction of roll film. Land's career as an inventor was launched in 1932 when he succeeded in aligning tiny crystals and embedding them in plastic to produce an inexpensive polarizer named Polaroid. He went on to amass over 500 patents for his inventions.

Part Six

ELECTRONICS
AND
COMPUTERS

265. Braun and the dawn of electronics. Ferdinand Braun stands before
us in the lecture hall of the Physical Institute at the University of
Strasbourg, where he was director. In 1897 he developed the oscilloscope,
or "Braun tube," to study high-frequency alternating currents. The principle of the Braun tube is fundamental to all television tubes today. Puzzled
as to why Guglielmo Marconi's original radiotelegraphy transmitter had
such a short range, Braun developed a sparkless antenna circuit, which
earned him a share (with Marconi) of the 1909 Nobel Prize in physics.

Electronics

Before electrons there were cathode rays. In the 1870s the English chemist and physicist William Crookes explored the discharges from a negative electrode in an evacuated glass bulb. What had been an interesting curiosity to Crookes and others was transformed into a practical instrument by the German physicist Ferdinand Braun, who in 1897 devised the cathode-ray oscilloscope—the basis of computer, television, and radar screens. Meanwhile, in the hands of Wilhelm Röntgen the "Crookes tube" had become the source of X rays. Physicians leaped at the possibilities presented by an instrument that would allow them to look inside the human body. Devising a practical instrument for actually seeing the shadow of the X ray was relatively short work for Thomas Edison, who discovered that crystals of calcium tungstate on a glass screen produced an effective "fluoroscope."

At just the time Edison was looking for the right chemicals for his fluoroscope, a much younger inventor-entrepreneur arrived in England, seeking help to exploit another new marvel of physics. Guglielmo Marconi was fascinated by the emanations created and identified almost a decade earlier by Heinrich Hertz in Germany. A number of researchers, such as Aleksandr Popov in Russia and Oliver Lodge in England, sought to use Hertz's electromagnetic waves to communicate. No one, however, brought to the task the drive and missionary zeal showed by Marconi. In 1901 he succeeded in detecting at his station in St. Johns, Newfoundland, the dot-dot-dot of the Morse code

letter S transmitted from a station in England. The economic possibilities of wireless seemed limitless.

As the economic potential became more visible, the incentives for technical improvement grew. In both transmission and reception technology the radio art advanced. In England, at the Marconi Company, John Ambrose Fleming devised in 1904 a new detector for wireless signals, making use of an effect that Thomas Edison had demonstrated some twenty years before. Edison showed that if an additional electrode were added to a light bulb, current would flow through the bulb's vacuum from the filament to the electrode, but not the other way. Fleming recognized that this Edison effect could be used as a "valve," converting high-frequency wireless signals into usable direct-current pulses. In modern parlance, Fleming's valve was a rectifier—the first true electronic device.

The diode, as Fleming's tube came to be called, was useful only as a detector. The real potential of electronic control became apparent in its successor, the triode of the American inventor Lee De Forest. By adding a third element, a "grid," between the filament and the electrode of the diode, De Forest produced in 1906 an effective amplifier and made long-distance wireless communication a truly practical technology. At first perceived only as a superior detector, his "audion" turned out to be a much-improved generator of radio-frequency signals. In fact, the device proved adaptable to a variety of techniques for controlling electronic signals, whether used in various combinations in ever more complex circuits or modified by clever tube designs—the "tetrodes," "pentodes," and other "-odes" and "-trons" that poured from corporate laboratories and small shops in the twenties and thirties.

The Marconi Company and its rivals successfully exploited the new technology for marine communications in the first decade of the twentieth century, but the true strategic significance of radio became clear only during World War I. Governments scrambled to take control over the increasingly complex web of corporate and patent interests that surrounded the technology. An entire generation of operators emerged from the war with new skills and interests, and, equipped with the inventions of Reginald Fessenden, E. F. W. Alexanderson, and Edwin H. Armstrong, radiotelephony (as opposed to wireless telegraphy) proved itself. The 1920s were thus ripe for the full exploitation of radio broadcasting, and by decade's end receivers were in millions of households in Europe and America.

Technological advancement in radio and allied fields followed a basic twentieth-century pattern. Achievements were often the product of both scientists and engineers, frequently working together in corporate laboratories. Major firms, such as General Electric, Marconi, Westinghouse, and AT&T, supported large full-time research establishments. Institutions like AT&T's Bell Laboratories became scientific centers of the first order, where Nobel Prizes were valued (and expected), just as were patents. Considering both government and corporate laboratories, the research facilities that supported the growth of electronic technologies were the most extensive ever seen.

Such resources made possible the extension of electronics capabilities between and during the wars. Television, the electron microscope, and radar were the most important products of the 1920s and 1930s, although their application was profoundly affected by World War II. While the war slowed down the development of some commercial technologies, it more than compensated by directing enormous resources toward the mastery of such phenomena as microwaves, ultrasonics, and solid-state electronics. The invention of microwave devices like the cavity magnetron and the klystron gave physicists and engineers control over a wide range of the electromagnetic spectrum, control that in the years after the war was extended to the molecular level with the maser and then to the realm of light itself with the laser.

It was indeed more and more at the molecular level that electronics found itself operating in the postwar years. The most spectacular evidence of this was the creation of advanced solid-state devices, especially the transistor. The phenomenon of semiconductivity, in which a material's resistance to electric current could be altered by such things as minute chemical impurities, temperature, light, and magnetic fields, had been observed in the nineteenth century, when the photoelectric properties of selenium attracted attention, as did the rectifying effect of galena (lead sulfide) crystals. In the first decades of the twentieth century, radio buffs looking for cheap rectifiers as detectors employed galena to build their "crystal sets," but little serious use of semiconductors was made until World War II. The need for particularly rugged radio and radar components led to the use of silicon and germanium in diodes, and postwar research sought to extend the applications of semiconductors further.

The transistor effect was discovered by the Bell Laboratories researchers John Bardeen, William Shockley, and Walter Brattain in December 1947. Their germanium point-contact transistor was the first solid-state device that could amplify signals, very much like De Forest's triode forty years before. There followed other, better transistors—junction and field-effect devices—and other materials—silicon and gallium arsenide. The reliability, small size, and low power requirements of the new amplifiers made them immediately attractive for many uses, such as in telephone relays and airborne radar. The real impact of semiconductors, however, came with the ability to manufacture them not as discrete components to be plugged into circuits but as integrated circuits. In the late 1950s researchers discovered that semiconductor materials could, with proper design, be made to serve all the functions of most circuits (resistors, capacitors, rectifiers, amplifiers). It was possible to manufacture on a single piece (or "chip") of, say, silicon all the components of a complex electronic circuit. In the early 1970s the art of integrated-circuit manufacture advanced far enough to put the basic circuitry of a digital computer on a single chip. The microprocessor thus brought the full power of semiconductor electronics to the service of the computer revolution, whose social and technical implications would shape the remaining years of the twentieth century.

Robert Friedel

266. Fleming's thermionic valve.
In 1883, Thomas Edison, working with a light bulb that had an extra metal plate in it, found a curious one-way flow of current to the plate. In 1904 the British electrical engineer John Ambrose Fleming patented a two-electrode tube, or diode, based on the Edison effect, for use as a detector in radiotelegraphy. Fleming said it "acted as a valve, permitting current to flow or to be shut off."

267. De Forest's audion. The output signals from Fleming's valve were weak. Seeking a means of amplification, the American inventor Lee De Forest in 1906 added to the tube a third electrode, which he called the grid. Placed between the filament and the plate, the grid could control the flow of current through the tube.

268. De Forest's amplifier. This extract from De Forest's notebook is dated August 28, 1912. In that year, while working as a research engineer with the Federal Telegraph Company in California, De Forest found that the three-electrode tube, or triode, could be used as an amplifier and an oscillator. In the drawing the plates are shown as rectangles, and the grids as wiggly lines. Here each tube has a pair of plates and grids. The audion (De Forest's name for his triode) provided superior detection of radio waves. Much of the subsequent work in electronics followed from it.

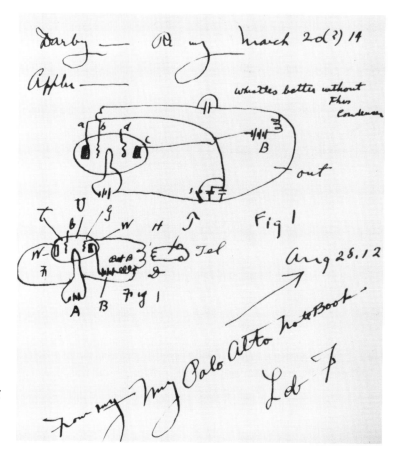

269. Marconi's receiving station in Newfoundland. The invention of radio involved the work of many men, but credit for the development of commercial wireless telegraphy is generally given to the Italian electrical engineer Guglielmo Marconi. In 1901, using a kite-borne antenna at St. Johns, Newfoundland, Marconi received transatlantic signals from Poldhu in Cornwall, England. This achievement produced a worldwide sensation. It proved that the curvature of the earth did not limit radio communication to 100 or 200 miles (as some experts had believed). The drawing shows the hospital at Signal Hill where Marconi established his western hemisphere receiving station.

270. De Forest's responder. Early wireless telegraphy experiments used a detector called a coherer. The British physicist Oliver Lodge showed in 1894 that iron filings in a tube would stick together, or "cohere," in the presence of radio waves and thus gave the device its name. Rather insensitive, the coherer had to be tapped, to "decohere" the filings, before another signal could be received. Among those who sought a better receiver was Lee De Forest. Early in the century, he developed the electrolytic "responder" shown here.

271. De Forest apparatus. De Forest was a prolific inventor but a poor businessman. His wireless company fell insolvent in 1906, and the De Forest Radio Telephone Company collapsed in 1910.

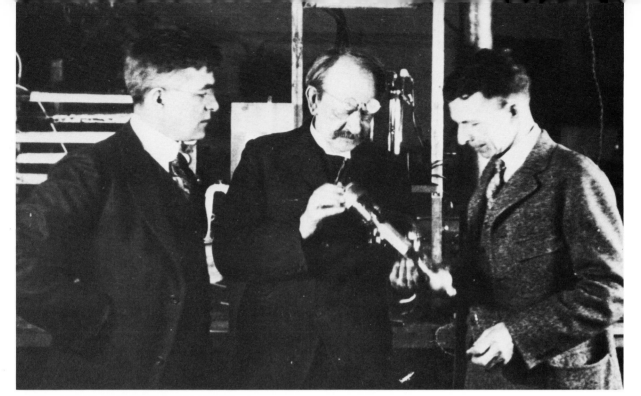

272. Langmuir, Thomson, and Coolidge. During World War I, J. J. Thomson (*center*), the discoverer of the electron, paid a visit to the General Electric Company research laboratory. Here he examines a pliotron tube, one form of multielement tube developed after De Forest's introduction of the audion. Thomson is flanked by chemist Irving Langmuir (*left*) and physicist William Coolidge. Two of the GE lab's most productive and famous scientists, Langmuir and Coolidge carried out work in electronics ranging from vacuum and X-ray tubes to the detection of submarines.

273. Early amateur radio broadcasting equipment. This apparatus, station 8XK, was used by the American electrical engineer Frank Conrad before the establishment of KDKA, the first commercial station, in Pittsburgh in 1920. Conrad helped set up KDKA, using a 100-watt transmitter, and made its first broadcast, relaying presidential election returns on November 2, 1920.

274. High-power tube. One of the largest radio tubes ever made was this water-cooled giant, the 250-kilowatt 320A developed in the late 1930s by Bell Laboratories as the power output for broadcasting stations. When in operation, the filament strand heated and expanded by half an inch, so jeweled bearings or guides had to be introduced. Eight 320As were installed in a 500-kilowatt station built in Mexico. Nevertheless, the market for such powerful tubes disappeared when the Federal Communications Commission limited broadcast power in the United States to 50 kilowatts.

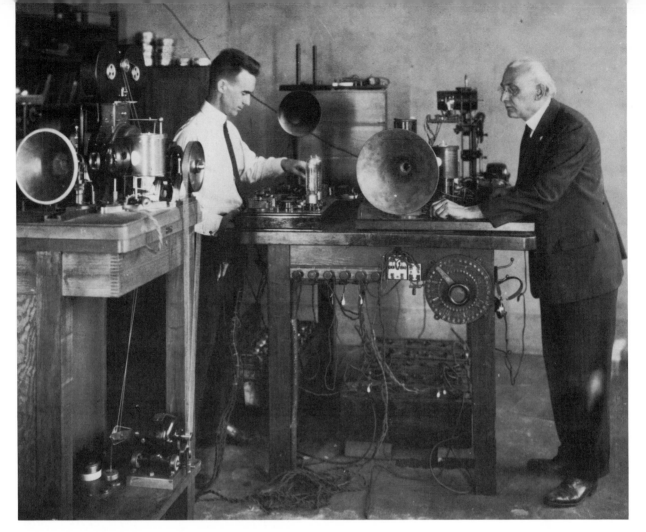

275. Hoxie and his Pallophotophone. The American inventor Charles Hoxie (*right*) is often regarded as the father of sound movies. He produced the first technique for converting sound to light and recording it on film. RCA's photophone sound, which was first used in 1928, developed largely from Hoxie's 1921 Pallophotophone, a sound-recording device using a mirror galvanometer.

276. Photomultiplier tube. The photoelectric effect—the emission of electrons from the surface of a substance when light falls on it—was discovered in the nineteenth century, but the invention of the electronic vacuum tube made it feasible to increase these faint currents. In the 1920s, "talkie" movies required a means to amplify a film's photographic sound track, thereby hastening the development of sensitive photoelectric devices. Here is an early photocell with electronic multiplication.

277. Zworykin and his iconoscope. The Russian émigré engineer Vladimir Zworykin is the father of modern television. While employed at the Westing-house Electric Corporation in the early 1920s, he developed the iconoscope and the kinescope. The iconoscope was a television camera tube based on the photoelectric effect. The kinescope was a television receiver. Together, they constituted the first all-electronic television system. Zworykin also developed a color television system in the 1920s.

278. First public demonstration of a waveguide. A waveguide is a hollow pipe or tube for conducting electromagnetic waves from one point to another. Here, a pioneer in the development of waveguides, the American physicist George Southworth (*left*), shows the device at a February 1938 meeting of the Institute of Radio Engineers. In this demonstration, Southworth connected one end of a flexible metal tube that was 50 feet, or 15 meters, long to a high-frequency source; measurements at the other end produced readings comparable to those at the source. The waveguide soon became an essential part of radar and other electronic systems.

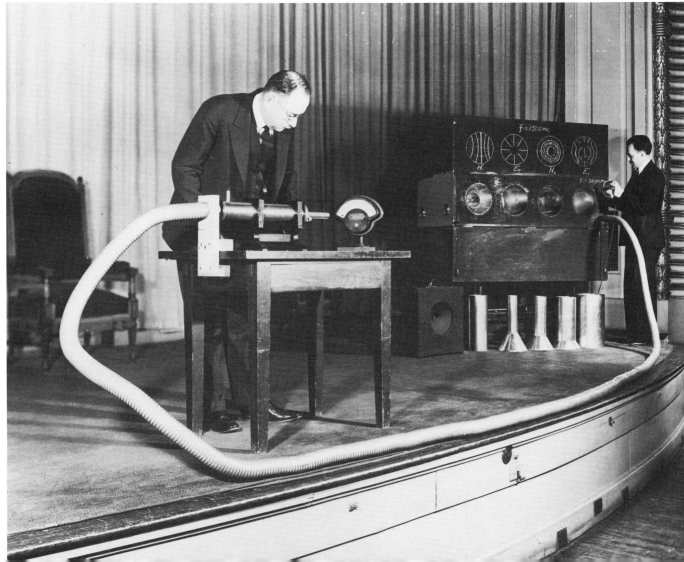

279. Ruska and his electron microscope. In an ordinary optical microscope, the specimen is illuminated with light, and the magnified image is seen after the light has been focused through lenses. To see objects smaller than light wavelengths, like viruses, it is necessary to use electrons with their much shorter wavelengths. The German physicist Ernst Ruska, with the help of Max Knoll, built the first electron microscope in 1931. An "emission" microscope, it employed magnetic fields as lenses to focus the electron beams and to produce an image of the electron-emitting cathode. In 1933, he built a transmission electron microscope in which the illuminating electrons passed through the specimen. The picture shows Ruska at the controls of the 1933 device.

280. High-voltage electron microscope. Specimens used in conventional transmission electron microscopes have to be ultrathin. Beginning in the 1960s researchers developed high-voltage electron microscopes permitting thicker specimens, a capability of particular value in materials science. The huge microscope we see here is at the University of Antwerp. In the large tank is a Cockcroft-Walton voltage multiplier that accelerates electrons to an energy of 1.25 million volts. Such energies permit examination of the atomic structure of thick samples.

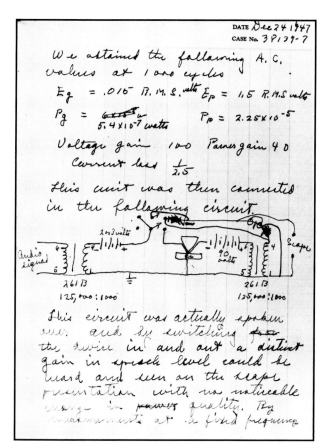

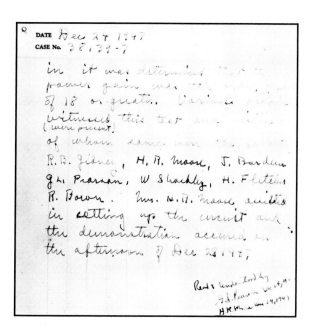

281. Invention of the transistor. This December 24, 1947, laboratory notebook entry, by Walter H. Brattain, records the events of the day before, when the transistor effect was first observed. Bell Laboratories scientists Brattain and John Bardeen, working under William Shockley, set up a circuit with what is now called a point-contact transistor to carry a spoken signal. "By switching the device in and out," writes Brattain, "a distinct gain in speech level could be heard and seen on the scope presentation with no noticeable change in quality."

282. First transistor (re-creation). For the first transistor, built at Bell Laboratories at the end of 1947, Bardeen and Brattain wrapped gold foil on the two sides of a small triangular plastic wedge, splitting the foil at the tip of the wedge and pressing it against a thin piece of germanium semiconductor. When a small current was applied to the germanium, a much larger current flowed from one foil edge to the other; in other words, the signal was amplified. The discovery of the transistor sparked a revolution in electronics; the triumph of miniaturization and semiconductors would transform society.

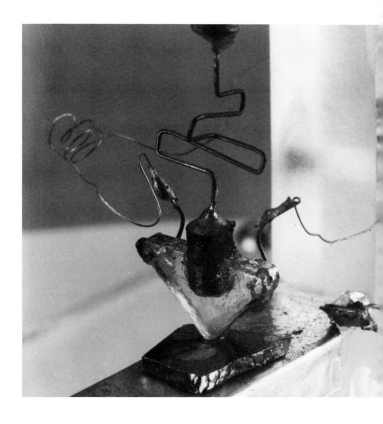

283. Junction transistor. The tricky point-contact transistor was quickly supplanted by the junction transistor, whose behavior was more predictable. Junction transistors sandwich one type of semiconductor around another. William Shockley built the first successful model in 1950. In the re-creation shown here, the junction transistor is the mudlike object standing a little more than a centimeter (½ inch) high in the middle of the board.

284. First integrated circuit. In 1958 the American engineer Jack Kilby assembled the first prototype of an integrated circuit—an electronic circuit formed not of discrete components but essentially of a single piece of material. A single tiny chip of germanium, smaller than 1½ millimeters (¹⁄₁₆ inch) square, was attached with wax to four electrical contacts. This first IC was a phase-shift oscillator, for converting direct current to alternating current. It functioned as a transistor, capacitor, and the equivalent of three resistors.

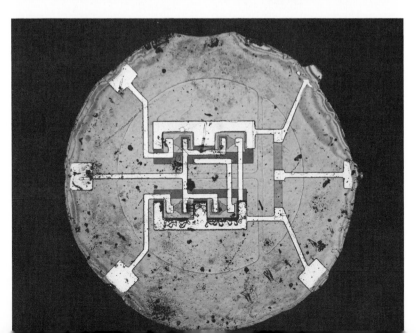

285. First planar integrated circuit. During the 1950s germanium gave way to silicon. In the planar process, the necessary impurities were diffused into selected regions of a silicon wafer through a mask so that the circuit's components lay in a single plane. Texas Instruments and Fairchild announced their first planar-process chips early in 1961. The Fairchild IC pictured here is a basic flip-flop logic chip, about 1½ millimeters in diameter.

286. First microprocessor. The Intel 4004, introduced in 1971, was the first commercial microprocessor. Combining arithmetic and logic functions on a single integrated circuit measuring 0.110 by 0.150 inches—about 3 by 4 millimeters—the 4004 was an entire computer central-processing unit in miniature. It had 2,300 transistors and could carry out 60,000 operations a second but could handle only four bits of data at a time.

287. 288K random-access memory. The development of memories on semiconductor chips allowed the miniaturization of computers, which had been using the bulky and expensive magnetic-core memory for internal storage. The first semiconductor memory of respectable capacity, 256 bits, was introduced in 1970. The entire chip was about the size of a single magnetic core, and it worked much more quickly than magnetic core memory. By 1981 IBM had developed this experimental memory chip; it provided 294,912 (or 288K) bits of "dynamic" random-access memory.

288. Integrated optical circuit. Because electrons have a certain mass and charge, there were limits to how far the miniaturization and improvement in speed of electronic devices could be carried. By the 1980s these limits were leading to increasing research on light as a means of transmitting information and, in an optical computer, of processing it. The integrated optical circuit shown here was used to switch light waves.

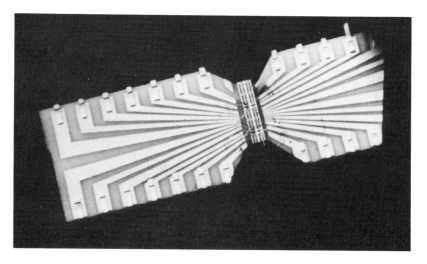

289. Programming ENIAC. Built at the University of Pennsylvania for the U.S. Army, the Electronic Numerical Integrator and Computor was the first electronic computer. ENIAC had more than 18,000 vacuum tubes and used 150 kilowatts of electricity. Setting the thousands of switches and plugging in the cables for the machine to perform a program could take two days. In this early 1946 photograph mathematician Arthur W. Burks and a programmer are checking a program.

18

*T*he *C*omputer *R*evolution

For ages astronomers were the principal consumers of numbers, but in the twentieth century other branches of science have become increasingly numerical. Startling advances in calculational technology have accompanied this growth, with the modern computer only the most recent in a long line of scientific instruments for processing information.

The desk calculator became a potent tool in the 1880s, when companies like Marchant and Burroughs introduced the first inexpensive and reliable calculators to meet the growing needs of large businesses. As the products improved over the next fifty years, they enjoyed increasing popularity among scientists. For example, the astronomer L. J. Comrie opened a calculating bureau in England that used desk machines to prepare mathematical tables and make astronomical and other scientific computations to order. Through the end of World War II it was not uncommon in American or European scientific institutions to find batteries of women, known as "computers," at work at desk calculators, each carrying out one step of the sequence of arithmetic operations that make up a complex scientific calculation.

A more powerful calculating technology, the punched-card tabulator, was developed to expedite analysis of the results of the 1890 U.S. census. Census data were recorded as holes punched in card stock. The machine read the data off the cards by passing electric current through the holes. Additional machinery was designed to count and sort the cards. The ability to store large amounts of data on the punched cards and to calculate and sort them automatically made this equipment attractive to scientists. It was used in the 1930s, for example, at the Columbia University Statistical Bureau to tabulate and analyze the results of educational and psychological tests, and at Iowa State University to solve problems of agricultural studies.

As science became more numerical, so did engineering. Special-purpose analog calculating devices helped meet the growing need for equipment that could handle engineering calculations. These devices used electrical circuits or mechanical linkages to model an engineering problem; the required numerical values were obtained by measuring electrical resistances or mechanical displacements. Many different analog calculators were introduced before computers made such devices obsolete in the 1950s. The most famous of these inventions were the differential analyzers designed by Vannevar Bush at the Massachusetts Institute of Technology (MIT) in the 1930s to solve differential equations that commonly arise in engineering.

The 1930s also saw efforts to develop faster digital calculating machines for complex scientific and engineering problems. Facing a year of work at a desk calculator for his dissertation research, physics student Howard Aiken designed a high-speed calculator whose electromechanical relays could carry out arithmetic operations hundreds of times more rapidly than any existing machine. His Harvard Mark I, installed during World War II, was used to make scientific tables and in other military and scientific applications. In the 1930s and 1940s Bell Laboratories developed a series of electromechanical calculators for its research scientists, employing the same relays used in telephone switching equipment. Europeans also built relay machines, notably the German inventor Konrad Zuse. Even this equipment was not fast enough for some purposes, so experimenters tried carrying out arithmetic operations electronically, with vacuum tubes. Two early successes were the Colossus machine, constructed by the British government to break German codes, and ENIAC, built for the U.S. Army at the University of Pennsylvania to calculate ballistics tables.

These electronic calculators had considerable raw computing power, but their uses were limited by the lack of a good way to control them. It sometimes took as long as two days to plug the cables and set the switches that controlled the sequence of arithmetic and logical operations for a calculation lasting less than an hour. This obstacle was overcome by the stored-program concept, invented at the University of Pennsylvania in 1945. Storing instructions as well as data inside the machine meant that problems could be set up more quickly and that the computer had greater flexibility because it could modify its instructions during the course of a computation. The first computers to use both electronic switching and stored programming appeared around 1950; among them were EDVAC, UNIVAC, Whirlwind, EDSAC, the Manchester Mark I, and the IAS machine at the Institute for Advanced Study in Princeton, New Jersey. These were one-of-a-kind, experimental machines built in government and university laboratories in England and the United States. Of extraordinary complexity, they could take five years to construct. The greatest problem was finding a low-cost device that could store information and provide fast access to any storage location. This was solved by the development of the magnetic core memory for the Whirlwind computer at MIT. The design fundamentals of computers became standardized in the 1950s, and commercial manufacture began. By the end of the decade commercial computers had all but replaced the experimental machines.

Users found that stored programming was not sufficient to harness the computer's power. Almost from the beginning, software was introduced to automate the programming process. Software tools, such as libraries of commonly used program routines, were developed in the early 1950s at Cambridge University and the Institute for Advanced Study. In the mid-1950s IBM and some of its large clients like General Motors began work on operating systems, which enabled the computer to manage its own administration (for example, allocating storage space to users or protecting the security of files). In the late 1950s the first programming languages, notably FORTRAN and COBOL, appeared. These allowed users to write their programs in languages resembling algebra and English rather than directly in the primitive code the machine understood.

While the fundamental design of computers did not change, the following decades brought radical improvements in components, logical design, and software, all of which made the computer significantly smaller, more powerful, less expensive, and easier to use. These changes occurred at an average rate of 30 percent a year. The biggest improvement was in switching components. The move from vacuum tubes to transistors in the late 1950s, to integrated circuits in the mid-1960s, to whole processing units on a single chip (microprocessors) in the early 1970s produced changes in every other aspect of computing technology from memory units to input-output equipment, to software.

The computer's impact on the sciences was enormous, qualitatively different from the effects of earlier calculating technologies. One of the first computers of the 1950s, that at the Institute for Advanced Study, was used for problems in numerical meteorology, the X-ray crystallography of proteins, stellar evolution, nuclear physics, and number theory. Other machines of the period were employed in petroleum exploration, aircraft design, and atomic weapons research. The speed and massive data-handling power of computers opened up new areas of scientific research, especially in the study of nonlinear phenomena like fluid dynamics. But the computer proved itself more than a supercalculator. It was used in entirely new ways. It could simulate unobservable processes like meltdowns of nuclear reactors, developments in the cores of evolving stars, or the long-term movement of tectonic plates. It could work in real time to control devices (like particle accelerators) too complex for humans to operate without assistance. Thus the world of the physical sciences came to be increasingly populated with these versatile machines.

William Aspray

290. Punched-card tabulator at the Census Office. The American inventor Herman Hollerith was the father of punched-card data processing. His electrical tabulating machine proved its worth in the 1890 U.S. census. It had taken seven and a half years to count the 1880 census results; complete analysis of the 1890 returns took only two and a half years. Hollerith's firm eventually developed into IBM.

291. Comrie. In the 1920s and 1930s the astronomer Leslie John Comrie opened the way for larger use of commercial calculating machines in science by showing how to "program" computational problems. He overhauled and mechanized the preparation of the *Nautical Almanac* in Britain and effectively laid the foundation for computational science in time for the advent of the electronic computer.

292. Columbia Statistical Bureau.
Established at the end of the 1920s, the bureau was equipped with a specially modified tabulating machine capable of automatically transferring numbers among its ten registers. Besides tabulating and analyzing the results of educational and psychological tests, this pioneering university statistical laboratory calculated inventories and astronomical tables.

293. Bush and his differential analyzer. Stymied in his efforts to solve certain difficult differential equations, the American engineer Vannevar Bush constructed a machine that provided solutions. Built at the Massachusetts Institute of Technology and completed in 1930, it was the first significant more or less general-purpose "automatic" analog computer. In the 1940s he built an electromechanical version using vacuum tubes.

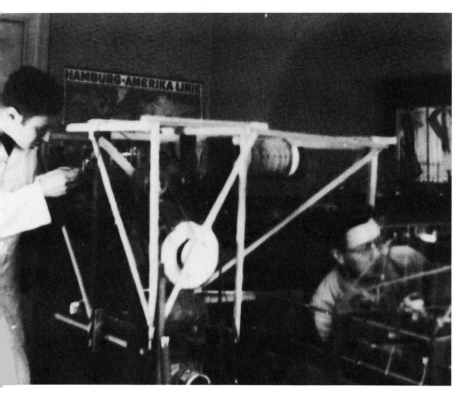

294. Zuse's mechanical computer.
The first fully functional program-controlled digital computer was built by the German engineer Konrad Zuse. It was completed in 1941 with the help of Helmut Schreyer; the machine had 2,600 electromagnetic relays and used the binary numbers with floating-point arithmetic. Zuse's first computer, a totally mechanical device built in 1938, is shown here. Schreyer is at left, and Zuse at right.

295. Atanasoff-Berry machine.
Completed in 1942 by John Vincent Atanasoff with the help of Clifford Berry at Iowa State College, this machine was a working electronic digital calculator, with such elements as a control console (*upper right*), a memory drum (*middle rear*), punched-card reader and punch (metal trays, *left*), and logic circuits (*bottom right*). An incomplete working prototype built in 1939 was the first machine to calculate with vacuum tubes.

296. Hopper and Harvard's Mark I. The Mark I, which went into operation in 1944, was the first large-scale automatic digital computer. This electomechanical computer was built by IBM to designs by Harvard professor Howard Aiken. Grace Hopper helped develop the operating programs for the machine and later became a leading figure in the design of computer languages, in 1951 conceiving the first compiler (a program that scans the programmer's instructions and generates a program in machine language understandable by the computer).

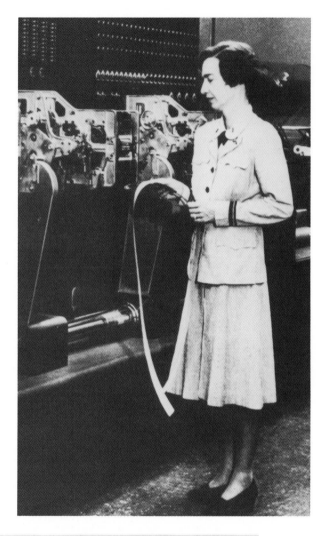

297. First real computer bug? In September 1945, Grace Hopper was working with Aiken's Mark II computer when the machine stopped because a moth was caught in a relay switch. Hopper and her Navy colleagues retrieved the bug and saved it in their logbook for posterity.

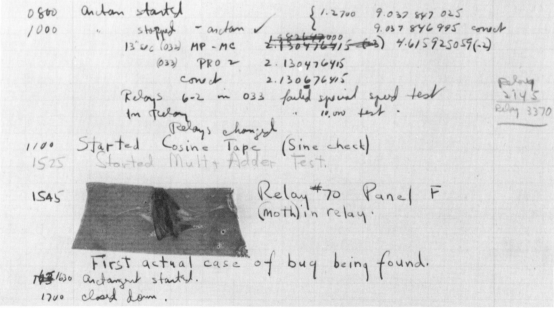

298. Colossus. As part of the British effort during World War II to crack German codes, a series of vacuum tube decoding machines, extremely fast for their time, were produced. The first Colossus went into operation at the end of 1943. In a sense, these were the first effective electronic digital computers, but they were special-purpose, code-breaking machines.

299. ENIAC: general view. This photograph of the electronic machine that really initiated the computer revolution was taken during ENIAC's dedication in February 1946. In the foreground are the two inventors: electrical engineer J. Presper Eckert, Jr., is turning a switch on a storage table, while physicist John Mauchly holds a control board.

300. Oppenheimer and von Neumann. Posing before the new computer at the Institute for Advanced Study in Princeton, New Jersey, at the machine's dedication in 1952 are J. Robert Oppenheimer (*left*), then the institute's director, and John von Neumann, whose seminal ideas on computer design included the stored program. Storing a computer's instructions internally—a concept first put forth in 1945 and later incorporated in the IAS machine—provided a powerful advance over the tradition of mechanical computers.

301. Manchester Mark I. The Mark I computer at the University of Manchester in England was the first fully electronic stored-program computer to operate. The prototype ran its first stored program in 1948, proving the validity of the new concept. The first commercial computer based on this machine, the Ferranti Mark I, was installed at the university's computing center in February 1951. In this 1949 report, the top picture shows builders T. M. Kilburn and F. C. Williams (*right*) standing in front of the prototype Mark I's control panel. Williams invented the machine's memory, which used cathode-ray tubes, one of which is seen in the bottom picture.

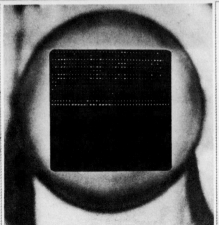

A CALCULATING MACHINE WITH A "MEMORY": THE CONTROL PANEL, AND A STORAGE TUBE IN USE.

WHEN Professor Geoffrey Jefferson delivered the Lister Oration at the Royal College of Surgeons of England on June 9, he disclosed that experiments were being conducted at Manchester University with a machine possessing a "memory." As a consequence, a number of misleading reports were published about the machine attributing to it almost human qualities. It is in fact similar to the American Electronic Numeral Integrator and Computer which we illustrated in our issue of November 9, 1946, but mainly differs from it in having a "memory," *i.e.*, it does not have to wait to be told what to do by a human operator when working out a problem, but, by the use of electronic circuits with a delay action, it is able to store a vast quantity of information which automatically takes its proper place in the calculation. This memory-storage system was invented by Professor F. C. Williams, who is seen in our photograph with Dr. Kilburn, feeding a mathematical problem into the machine for solution. The Manchester Automatic Sequence-Controlled Calculating Machine has been devised and constructed to undertake a wide variety of complex calculations which would take human *Continued opposite.*

(ABOVE:) THE CONTROL PANEL OF THE AUTOMATIC SEQUENCE-CONTROLLED CALCULATING MACHINE, AT MANCHESTER UNIVERSITY, SHOWING THE MONITOR CATHODE-RAY TUBE WITH DR. T. KILBURN (ON LEFT) AND PROFESSOR F. C. WILLIAMS, INVENTOR OF THE MEMORY-STORAGE SYSTEM EMPLOYED (ON RIGHT), FEEDING A PROBLEM INTO THE MACHINE. (LEFT:) ONE OF THE CATHODE-RAY STORAGE TUBES, SHOWING THE POINTS OF LIGHT INDICATING A MATHEMATICAL PROBLEM IS BEING SOLVED.

Continued.

beings, using ordinary methods, possibly months to carry out, where the machine takes only an hour or so. The human controller has to decide how the Calculating Machine can perform the desired calculation, and draws up a list of "instructions" for it to obey. He breaks up the complex calculation into a series of simple basic operations and translates these from numbers into a specified code. For instance, the operation of subtracting one number from another is called operation No. 29 in the code. The list of "instructions" is fed into the machine, and the initial numbers (in code) on which it is to operate are then loaded into a special position. All the information having been fed into the Calculating Machine, its "memory" can be switched on to start operations. When the machine has worked out the whole problem, a red light switches on and it stops automatically. The final result can then be read off the monitor cathode-ray tube (shown in our photograph) in the form of light dots which are translated into figures by the human controller. A photograph of the complete apparatus appears elsewhere in this issue.

302. Whirlwind. Project Whirlwind at the Massachusetts Institute of Technology began as an analog aircraft simulator but developed into a huge digital computer. It went into routine operation in 1951. Designed for real-time applications, it was the fastest computer of its time and pioneered the use of magnetic-core memory.

303. Magnetic-core memory. Each little ring, or "core," in the array is made of a ceramic ferrite material and represents a single bit of information. Depending on the direction in which the core is magnetized, it can code a zero or a one.

304. Forrester. Jay Forrester, an electrical engineer who headed Project Whirlwind at MIT and invented the core memory system, holds a single magnetic-core memory plane of the type that was used in the Whirlwind machine.

305. UNIVAC. The first Universal Automatic Computer, or UNIVAC, the first commercially available computer in America, was delivered to the U.S. Census Bureau in June 1951. A UNIVAC was used by CBS-TV in 1952 to predict the results of the presidential election on election night. Here, J. Presper Eckert, Jr., one of the machine's developers, explains the system to CBS newsman Walter Cronkite.

306. EDSAC. Making its first calculation in May 1949, the Electronic Delay Storage Automatic Computer at Cambridge University was the first fully functional modern computer—an electronic stored-program machine with substantial computational ability. (The Mark I computer at Manchester was only a prototype machine.) EDSAC used a mercury delay line for its memory.

307. IBM 704. An early IBM scientific computer with core memory, the 704 was introduced in 1954. The first relatively simple programming language, FORTRAN (for "formula translation"), was initially developed for this machine. Introduced in 1957, FORTRAN was eventually recognized as a landmark computer language.

308. IBM 7090. The first computers to use transistors instead of vacuum tubes went on the market in the late 1950s. A member of this second generation of computers, the IBM 7090 was dominant in the scientific computer market between 1960 and 1964. With the transistor, computers could be smaller, faster, more reliable, and more powerful.

Part Seven

COSMIC VISTAS

309. Arecibo radio observatory. This huge radio telescope near Arecibo, Puerto Rico, was built in the early 1960s by Cornell University. It is 1,000 feet, or 305 meters, in diameter. The main reflector is a wire mesh suspended in a natural hollow in the earth; a movable receiver is suspended above it on cables. The facility can be used actively or passively. For example, it can bounce radar waves off Venus in order to map the planet's surface, or it may simply record incoming signals, such as the intense radio bursts from pulsars.

19

*R*adar and *R*adio *A*stronomy

The use of radio waves to detect and locate objects was nurtured by the military advantages of radar, but there were scientific benefits as well. Radio waves provided a new way of exploring the universe. As early as 1894, just six years after radio waves had first been produced in the laboratory, the British physicist Oliver Lodge initiated a vain attempt to observe solar radio emission.

Extraterrestrial radio waves were first detected in 1931 by Karl Jansky at the Bell Telephone Laboratories in New Jersey. Jansky was studying atmospheric "static" responsible for interference in telephone communications. His rotatable antenna found natural radio noise from local thunderstorms, distant thunderstorms, and a source associated with the center of the Milky Way. Jansky later showed that the extraterrestrial radiation came from along the plane of the Milky Way, with the greatest intensity at the galactic center.

This work was followed up on by the American radio engineer Grote Reber, who in 1937 built a parabolic reflector at his home in Wheaton, Illinois. Reber observed at night, since by day he had a job to hold down. He tried several wavelengths before his persistence paid off: at a wavelength of 1.9 meters he made the first maps of the Milky Way. Reber discovered peaks of radiation in the constellations Sagittarius, Cygnus, and Cassiopeia. He had in fact picked up radio waves from the nucleus of our galaxy, from a remote radio galaxy, and from the remnant of an exploded star.

Meanwhile, work was proceeding apace on ra-

dar, which uses radio waves reflected, rather than emitted, by an object. In 1900 the Croatian-born American inventor Nikola Tesla had suggested using electromagnetic waves to "determine the relative position or course of a moving object . . . the distance traveled by same or its speed." The first practical radar systems appeared in the 1930s. The invention by British researchers around 1940 of the microwave generator known as the cavity magnetron made possible microwave radar (more sensitive than radar using longer wavelengths) of high power. Intimations of the possible value of radar to astronomy had come already in the 1930s: radio transmissions sometimes seemed to be reflected from meteor trails, a property that later played a key role in the study of meteors.

In 1942 the British physicist J. S. Hey, while working on radar, accidentally detected radio emission from the sun. After World War II the talented radio physicists Martin Ryle in Cambridge and Bernard Lovell in Manchester made the quantum leap from the radar technology of war to the peacetime exploration of the cosmos through radio waves. They initially used radar equipment salvaged from the military but soon turned to specially built radio telescopes. Their reflectors of sheet iron and mesh did the same job as reflecting mirrors in optical telescopes.

In 1946 natural thermal radiation from the moon was detected by Robert Dicke in the United States, and the U.S. Army Signal Corps obtained radar echoes from the moon. Radar echoes later provided one of the main means of mapping the surface of Venus. This planet has an immensely thick atmosphere, but radar can penetrate the clouds and has shown the existence of mountains, valleys, and plains.

The first great theoretical breakthrough in radio astronomy came from the Netherlands in 1944. Prompted by Reber's publications, Jan Oort of the Leiden Observatory suggested to the young astronomer H. C. van de Hulst that he look for ways in which spectral lines of radio emission might be produced. Van de Hulst found that atomic hydrogen should yield a detectable signal with a wavelength of 21 centimeters. In 1951 radio scientists at Harvard University detected this 21-centimeter line, and the signal was later used to map the spiral structure of galaxies, including our own. Much of what we know about the behavior of interstellar matter comes from studies of the 21-centimeter line.

From the start the pioneers of radio astronomy had to cope with intrinsically faint signals. All the radio energy ever collected by all the telescopes in the world would scarcely lift the book you are reading more than a few millimeters from the floor! Radio astronomers also had to solve severe problems of resolution. Radio wavelengths are about a million times longer than light wavelengths. In practical terms this means that a radio telescope needs to be many kilometers in diameter before it can achieve a sharpness of detail matching the photographs from optical telescopes.

Lovell pioneered the construction of large single-dish telescopes. His 250-foot instrument at Jodrell Bank commenced operations in 1957, and it is still one of the largest fully steerable radio telescopes in the world. Fixed dishes can be made much larger than steerable ones. Near Arecibo in Puerto Rico a static 1,000-foot telescope was constructed in a natural crater in the ground in the early 1960s; it remains to this day the world's largest dish antenna.

The problem of resolution was solved by Ryle in Cambridge and by J. L. Pawsey's group in Sydney, Australia. They made radio analogues of A. A. Michelson's stellar interferometer. By linking pairs of small radio telescopes, they could synthesize some of the properties of large dish telescopes, particularly the ability to map distributed radio sources in fine detail. The Very Large Array in New Mexico today can map objects in the radio spectrum with detail exceeding that of most optical telescopes. The highest resolution, far greater than that available with any optical instrument, is ob-

tained by linking radio telescopes on different continents in a technique known as very-long-baseline interferometry.

Radio astronomers have made discoveries of immense importance to astronomy as a whole. Radio waves from the sun are monitored at observatories dedicated to that purpose. Another powerful radio transmitter in the solar system is Jupiter, its radio waves being triggered by the movement of its satellite Io through the planet's strong magnetic field. Within our galaxy radio astronomers have mapped the spiral structure of interstellar matter and probed into the heart of the Milky Way. In 1967, Antony Hewish and colleagues at Cambridge University announced the discovery of pulsars, rapidly bleeping radio sources now known to be neutron stars just a few miles—say 10 kilometers or so—in diameter.

Beyond the Milky Way radio astronomers survey the neutral hydrogen in normal galaxies such as the Andromeda nebula. But the richest prize in deep space is the intense radiation from radio galaxies and quasars. About one galaxy in a million is a powerful radio source, the prototype being Cygnus A, discovered by Hey in 1946. Related to radio galaxies are the even more powerful quasars, which emit extremely intense light.

Radio astronomy plays a crucial role in cosmology, the branch of astronomy that considers the universe as a whole. Early on in radio astronomy Ryle tried to test models of the universe by counting the distribution of radio sources on the sky. This led to a dispute between Ryle, who felt that radio-source counts supported the Big Bang theory of the origin of the universe, and Fred Hoyle (also of Cambridge), who advocated a steady-state theory. The furious argument led to vigorous funding of radio astronomy in Great Britain. A decisive discovery, however, came from an unexpected quarter. Just as Jansky had accidentally detected cosmic radio waves while studying static, so another investigation related to communications technology led, in 1965, to a golden moment in cosmology. Arno Penzias and Robert Wilson of the Bell Telephone Laboratories detected a faint background hiss in the microwave region of the spectrum. Found to come from the universe at large rather than from any discrete sources, it was rapidly accepted to be fossil radiation from the primeval fireball of the Big Bang.

Simon Mitton

310. Antenna for the first complete U.S. radar. "Topsy," an antenna erected above a building at the Naval Research Laboratory in Washington, D.C., in the late 1930s, could rotate to scan in any direction. Although the word "radar" (an acronym for radio detection and ranging) was coined during World War II, radar's beginnings date much farther back. In 1904 the German engineer Christian Hülsmeyer patented a method for using radio waves to detect obstacles.

311. Cavity magnetron. A tube for generating microwave power, fundamental to the success of radar, the cavity magnetron was developed in Great Britain around 1940. Here we see a 1943 scene at the Radiation Laboratory at the Massachusetts Institute of Technology. The British physicist E. G. Bowen (*seated*), who had brought a magnetron to the United States, is shown an American-made copy by director Lee DuBridge (*left*) and physicist I. I. Rabi.

312. Antiaircraft radar. The SCR 268, photographed in use in Italy in 1944, was a spotlight-type detector used to direct antiaircraft fire. Its 150-centimeter wavelength was the shortest feasible before the cavity magnetron was introduced. "SCR" was an acronym for "Signal Corps Radar."

313. Project Diana. Radar pulses were first bounced off the moon in 1946, from this U.S. Army antenna at the Signal Corps Engineering Laboratory at Fort Monmouth, New Jersey. Although primarily a stunt, the achievement opened up a new method for the study of the propagation of radio waves through the atmosphere, and it paved the way for later precision measurements of planetary distances.

314. First radio telescope. Working at the Bell Telephone Laboratories in Holmdel, New Jersey, to investigate static interfering with radiotelephone communications, Karl Jansky built the rotating antenna at right. In 1931 he discovered that a major form of interference was coming from the sky. The primary source, he soon concluded, was in the constellation Sagittarius, where the center of the Milky Way galaxy was located. Radio astronomy was almost stillborn, however, because Jansky's supervisors in the laboratory decided they could do nothing about cosmic interference and so reassigned him elsewhere.

315. Reber's parabolic antenna. Eventually the engineer Grote Reber carried on Jansky's work. In the backyard of his home in Wheaton, Illinois, Reber in 1937 constructed a sheet-iron, dish-shaped receiver that was 31 feet (9½ meters) in diameter. With this antenna he made the first contour radio maps of the entire northern sky. A lone pioneer, his farsighted work was not improved on for over a decade.

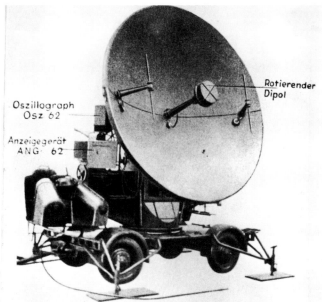

316. A war surplus bonanza for radio astronomy. In the late 1940s, thanks to the availability of high-quality captured German radar antennas and receivers, a number of Allied scientists who developed radar in World War II became radio astronomers. Somewhat larger versions of the "Würzburg C" radar above played key roles in England, the Netherlands, the United States, France, and Sweden.

317. Discovery of neutral hydrogen in space. Using this horn antenna, the Harvard University physicist Edward Purcell and his graduate student Harold Ewen (shown above) in 1951 detected the microwave radiation of 21-centimeter wavelength emitted by atomic hydrogen in interstellar space. With the 21-centimeter radio signals, whose existence had been predicted in 1944, scientists could find hydrogen clouds for use in measuring the rotation of galaxies and mapping the Milky Way's spiral form.

318. Spiral arms of the Milky Way. Prolonged efforts by optical astronomers to discover the Milky Way's spiral arms finally bore fruit in 1951. Comparing the observed distribution of hydrogen emission nebulosities in the Milky Way with the pattern in the Andromeda galaxy (superimposed on the left), where the spiral arms were clearly visible, led astronomers at Yerkes Observatory—W. W. Morgan and his collaborators—to detect the spiral arm mapped here.

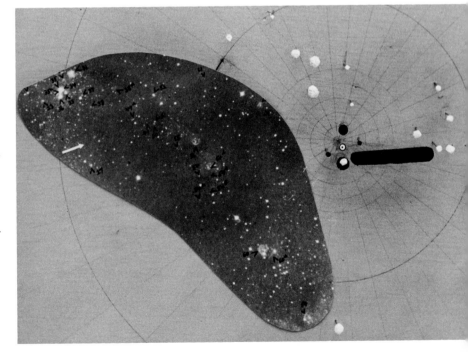

319. Radio mapping the Milky Way. After the discovery of the 21-centimeter radio waves from hydrogen, astronomers at Leiden Observatory in the Netherlands immediately began using them to survey the northern part of the Milky Way galaxy. The resulting map of hydrogen distribution, reproduced here, was published in 1957. The regions of high hydrogen density correspond to our galaxy's spiral arms. Distances from the center of the Milky Way are in thousands of parsecs. A similar survey at Sydney, Australia, for the southern hemisphere was completed in 1959.

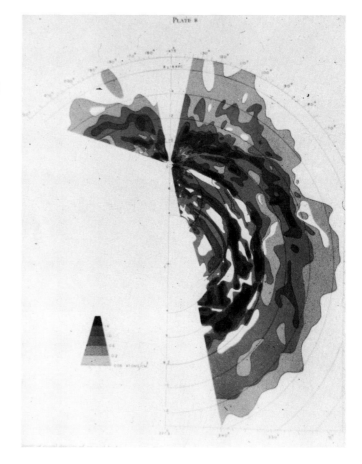

320. Westerbork Synthesis Radio Telescope. This interferometer array of fourteen steerable antennas at Westerbork in the Netherlands went into operation in 1970. Each antenna is 25 meters (80 feet) in diameter. The antennas are arranged in a row 3 kilometers (about 2 miles) long. Working together, the network can yield a resolution equivalent to that of a single antenna 3 kilometers in diameter.

321. Mark I dish at Jodrell Bank.
From its humble origin in 1945 with a mobile army radar, Bernard Lovell's radio astronomy observatory near Manchester quickly developed into England's major astronomical observing facility. Construction of an unprecedentedly big steerable antenna—250 feet, or 76 meters, across—began in 1952 and was completed in 1957. The Mark I dish immediately went to work tracking the carrier rocket of Sputnik, the first artificial earth satellite. This 1958 view is from the control room.

322. Solar radio observatory.
This installation for studying the radio activity of the sun is located at Culgoora, Australia, 300 miles northwest of Sydney. The complex system, which went into operation in 1967, has 96 antennas, each 13 meters (43 feet) in diameter, arranged in a circle 3 kilometers across.

323. Very Large Array. The three-armed interferometric array known as the VLA is located about 50 miles west of Socorro, New Mexico. Dedicated in 1980, the VLA has twenty-seven antennas, each 25 meters (82 feet) across. The antennas, which move on railway tracks, are here shown in their most compact configuration. At maximum extent, the arms extend up to 21 kilometers (13 miles).

324. Cambridge fixed array. The length of Martin Ryle's early interferometric arrays at Cambridge University enabled them to pinpoint radio sources in the sky with considerable precision, thereby giving great impetus to optical identifications. A succession of standard catalogs resulted. Here we see the so-called "4C" array in a 1968 photograph.

325. Goddard. The American physicist and engineer Robert Goddard was one of three scientific pioneers of rocketry. The other two were the Russian Konstantin Tsiolkovsky and the German Hermann Oberth. Goddard designed, built, and flew the first liquid-propellant rocket. The flight took place on March 16, 1926, near Auburn, Massachusetts. The rocket rose to an altitude of 41 feet (just over 12 meters) and landed 184 feet away.

CHAPTER

20

*M*en on the *M*oon

On December 19, 1972, at 2:25 P.M. local time, Apollo 17 plunged into the Pacific Ocean, bringing back to earth astronauts Ron Evans, Gene Cernan, and Harrison ("Jack") Schmitt, a geologist and the only "scientist-astronaut" to touch the lunar surface. Thus ended the era when human beings walked on the moon. A dream older than manned flight itself had been fulfilled.

The Apollo program brought back 850 pounds, or about 385 kilograms, of lunar material, over 16,000 photographs, and a mass of physiological data and sensory information from the lunar environment. It was a result of a complex mix of national imperatives and Cold War drives, blended with the ambitions of a small band of military and industry-based rocketeers who looked back to the visions and dreams of three men: the Russian Konstantin Eduardovich Tsiolkovsky, the German Hermann Oberth, and the American Robert Goddard.

Tsiolkovsky's articulation of a rocket-powered spaceship in 1903 may have set the stage for the visionaries, but the achievement of heavier-than-air flight by the Wright brothers in the very same year set the stage for the first practical steps in air and space flight. Whether or not the Wright brothers' powered flight was the first ever (a point questioned by some), it drew the world's attention to the possibilities of flying machines. It was not a stunt. Carefully planned, from drafting table to machine shop to test gliders to powered flight, it was one step in a continuum of systematic work

that made controlled manned flight practical. Worldwide reaction to the Wright brothers' feat was swift—within only a few years aircraft of many types appeared.

Reaction to the rocketry experiments of Robert Goddard, by contrast, was agonizingly slow. Goddard, who began his research before World War I, envisioned rocketry as "a method of achieving extreme altitudes" and of ultimately reaching the moon. His systematic but underfunded research between the wars yielded slowly but steadily improved rocket designs. Goddard obtained patents for combustion chambers, pressurized fuel systems, nozzles, and other elements of liquid-fueled rockets that could provide controlled propulsion in the vacuum of outer space. Virtually ignored in the United States, he had a far greater influence in Germany. The noted German engineer Wernher von Braun was later to say that many of Goddard's designs formed the basis for all modern rocket engines.

From the early 1930s, rocketry flourished in Germany. Limited by the Treaty of Versailles at the end of World War I to only small-caliber artillery, the German Army turned to rockets as it prepared for the next war, exploiting a highly developed popular interest embodied in the German Verein für Raumschiffahrt (Society for Space Travel). Members of the society were recruited to help develop military rocketry. By the end of 1932 a young Wernher von Braun was at work, and by the end of the 1930s liquid-fueled engines with multiton thrusts were undergoing static tests in what was a carefully orchestrated program of systematic development. The result was the German A-4 missile, also known as the V-2, first successfully test launched in 1942 and directed toward enemy targets in 1944.

The damage caused by the V-2 was more psychological than physical, but its existence and potential fostered a belief in both the United States and the Soviet Union that ballistic missiles equipped with atomic warheads would be the weapons of the next war. As World War II ended, the two superpowers captured V-2 parts and personnel for use in their own rocket development programs. Von Braun, for example, went to the United States.

Instead of TNT the V-2s now carried aloft scientific experiments in order to obtain new knowledge about the upper atmosphere and near space: the new battlefield. Experimenters in several laboratories prepared spectrographs, cosmic ray detectors, and other devices to probe the upper atmosphere and the solar spectrum. They were as interested in problems in physics, astronomy, and the geosciences as their military patrons were in understanding the upper atmosphere's influence on long-range radio communications and ballistic flight. While the military tested the V-2 rockets and learned how to build better ones, partly guided by Germans such as von Braun, a new cadre of scientists took shape that was skilled in conducting experiments in near space.

In a global enterprise called the International Geophysical Year, the United States and the Soviet Union each planned scientific earth satellite programs. The Soviets were first to make theirs a reality, launching Sputnik on October 4, 1957. The fledgling Vanguard was left on the launchpad.

Sputnik certainly was a surprise to Public America. The National Aeronautics and Space Administration (NASA) was formed in short order, and an ad hoc space race was on. American public policy demanded not only prowess but superiority in space technology. The United States had to catch up and pass the perceived Soviet threat from space. Any reasonable ballistic missile delivery system was likely to be automatic and computer controlled, but the visible demonstration of prowess had to be on a peaceful playing field. By the early 1960s strategic planners in the White House and NASA knew that the United States had a better than even chance to send a man to the moon and return him safely. The systematics of the Wright brothers and of Robert Goddard took second place to the political imperative of scoring the first touchdown on the moon.

To be sure, systematic steps were taken to develop the transport capability. These included both unmanned (Ranger and Surveyor) and manned (Mercury and Gemini) forays into space. And of course the Apollo program did return an enormous amount of excellent scientific information from the moon. Analysis of Apollo lunar rock samples provided ages and compositions that refined and confirmed the automated Surveyor's findings about the lunar surface. Above all, thanks to Apollo, mankind finally touched another celestial body.

Afterward, NASA salvaged some leftover Apollo hardware for the Skylab solar observatory and for the Apollo-Soyuz Test Project, which saw a U.S. craft and a Soviet craft join in orbit in 1975. But on the whole, American scientific research in space was carried on through a series of unmanned satellites and interplanetary space probes. However, by the end of the 1970s a severely curtailed American space program was focused on the reusable manned low-earth-orbit system called the shuttle, whose name derived from its original purpose to "shuttle" crews and equipment to a space station. Meanwhile, the Soviet program continued with a blend of expendable launch vehicles for both manned and unmanned flight. The Soviets' Salyut series of space stations, which began in 1971, was followed by the Mir (launched in 1986), permitting a continuous manned presence in space. In this manner the international legacy of the Wright brothers is kept alive.

David DeVorkin

326. Wright brothers' first manned flight. The American inventors Wilbur and Orville Wright are generally considered to have designed and built the first aircraft to perform powered free flight while carrying a passenger. This first heavier-than-air flight took place on December 17, 1903, at Kitty Hawk, North Carolina, with Orville Wright in the plane.

328. V-2. A landmark in the advance of rocketry, developed during World War II, was the German V-2 (Vergeltungswaffen Zwei, or Vengeance Weapon Two). Burning alcohol and liquid oxygen, the V-2 was the first rocket to go faster than the speed of sound. It carried a 1-ton warhead and had a maximum range of about 350 kilometers (220 miles). After the war, captured V-2s were used extensively in Soviet and American research.

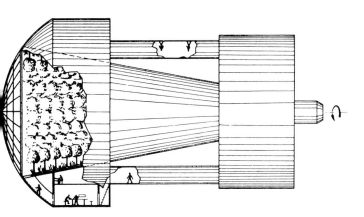

327. A Tsiolkovsky spaceship design. While working as a mathematics teacher in provincial Russia in the late nineteenth century, Konstantin Tsiolkovsky began his path-breaking research in theoretical astronautics. He studied the possibility of interplanetary flight, expounding on rockets and the liquid-propellant engine in "Investigation of Interplanetary Space by Means of Rocket Devices" (1903) and subsequent works.

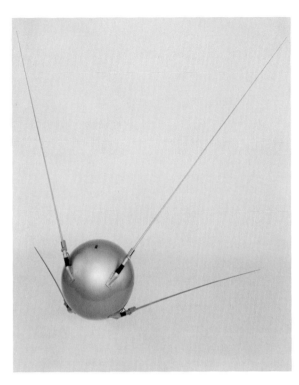

329. Sputnik. The Space Age began when the Soviet Union launched the first artificial earth satellite, Sputnik, on October 4, 1957. This spherical instrument package weighed 84 kilograms (185 pounds). Circling the earth at a maximum distance of 946 kilometers (588 miles) and a minimum of 227 kilometers, it took ninety-six minutes to complete an orbit. A scant month after the first triumph came an even greater achievement: the 509-kilogram (1,120-pound) Sputnik 2 carried the dog Laika into orbit. Sputnik fell back into the earth's atmosphere and disintegrated in early 1958, and Sputnik 2 followed a few months later.

330. Gagarin. The first human to travel into space was the Soviet cosmonaut Yuri Alekseevich Gagarin. On April 12, 1961, he rode the 5-ton Vostok 1 spacecraft into orbit, circled the earth once (reaching a maximum altitude of 327 kilometers, or 203 miles), and parachuted safely back to earth. In the photograph Gagarin, dressed in his cosmonaut suit, is waving good-bye before entering the Vostok. (In 1968 Gagarin died in the crash of a jet aircraft on a routine training flight.) On August 6, 1961, a second cosmonaut, German Stepanovich Titov, completed seventeen orbits aboard the Vostok 2 before parachuting back to the U.S.S.R.

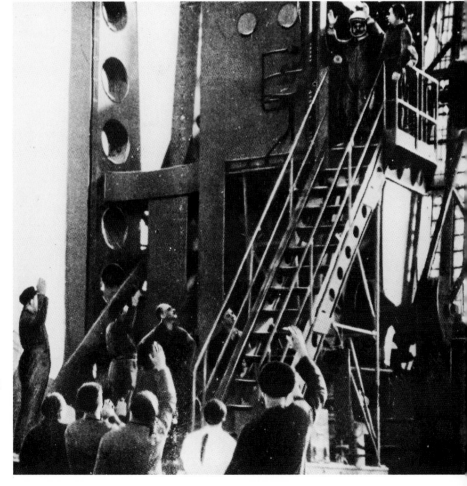

331. Soyuz 19. This Soviet spacecraft, launched on July 15, 1975, docked with the American spacecraft Apollo for two days while in orbit. The photograph was taken from a window of the Apollo. The first in the Soyuz series was launched in April 1967 and completed seventeen orbits. Its pilot, Vladimir Mikhailovich Komarov, was killed during reentry, becoming the first person known to have died during a space mission, although three American astronauts had lost their lives in a disastrous test pad fire a few months earlier.

332. Space station Mir. Early in 1969 the Soviet spacecraft Soyuz 4 and Soyuz 5 joined and docked in orbit, forming what might be called the first space station. More substantial were the stations in the Salyut series, which saw its first launch in 1971. Another advance was the Mir, sent into earth orbit in February 1986. Designed as a "base unit," the Mir could accommodate specialized laboratory modules.

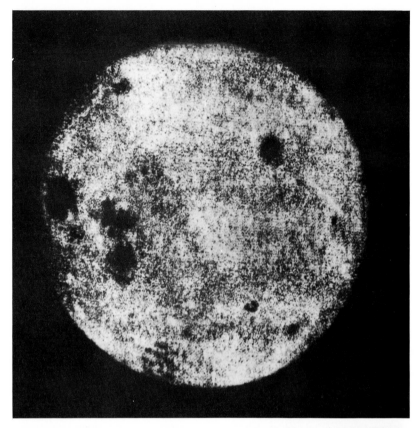

333. Far side of the moon. Since the moon always presents the same side toward the earth, its backside remained unknown until the Space Age. The Soviet space probe Luna 3 provided the first view of the moon's other face during its forty-minute flyby in October 1959. Despite the crudeness of the images, Soviet selenographers complied an atlas of 498 features. Reproduced here is the best global picture, reconstructed with American image processing and enhancement. Perhaps the most surprising discovery was that the moon's backside lacked large maria, the vast dark basaltic plains that form the features of the familar "Man in the Moon" visible from earth.

334. Ranger 7 moon montage. In July 1964 the U.S. lunar probe Ranger 7 sent back more than 4,300 televised pictures, giving the first high-definition close-up views of the moon's surface. This full frame was made from an altitude of about 5 kilometers (3 miles) just 2.3 seconds before the fatal impact on Mare Nubium. The smallest inset is 50 meters (164 feet) across and shows craters just a few feet in diameter.

335. Rolling stones on the moon.
This remarkably detailed image of
the tracks of two large boulders that
had rolled down from the central
peaks of crater Vitello is one of
many produced by the American
probe Lunar Orbiter 5 in August
1967. The larger track, about 25
meters (82 feet) wide, shows a re-
petitive pattern created by the rolling of an irregularly shaped mass.
Unlike the Ranger probes, which
crash-landed, the Lunar Orbiters
circled the moon repeatedly in their
surveys.

336. Lunar landscape with rilles. The crater Prinz
and its associated pattern of rilles were recorded from
Lunar Orbiter 5 on August 18, 1967. The rilles, which
look curiously like meandering stream beds, are actually subsidence trenches in a particularly deep section
of the outer fragmented stony layer of this waterless
plain. In 1966 and 1967 the Lunar Orbiter series provided a television reconnaissance of the entire moon,
including backside portions omitted in the earlier
Soviet views; Orbiter 5 examined the areas under
consideration for manned landings.

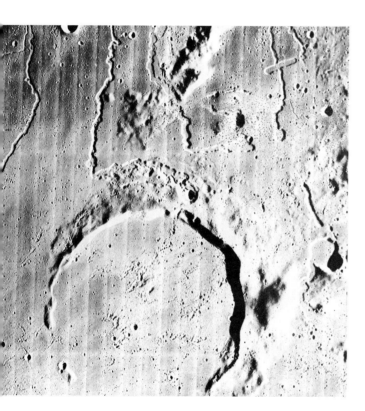

337. Mare Orientale. Enormous mountain ranges
6,000 meters (20,000 feet) high circle the dark central
plain. These mountain rings are 600 and 1,000 kilometers (about 400 and 600 miles) across. Although this
striking impact feature near the western limb of the
moon had been partially visible from the earth, its true
appearance was revealed only in pictures taken from
lunar orbit. The television view shown here was sent
back by Orbiter 4 on May 25, 1967.

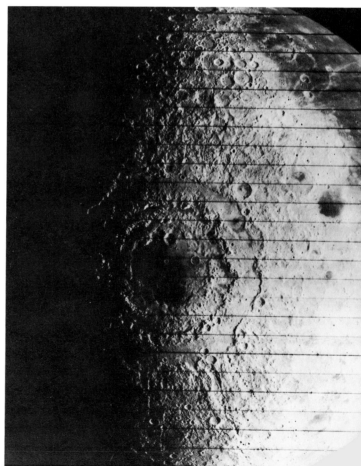

338. Manned science on the moon. The first men to walk on the moon were carried there by the U.S. spacecraft Apollo 11 in July 1969. Here we see astronaut Edwin E. Aldrin, Jr., unfurling an aluminum foil sheet intended to catch particles of the solar wind issuing from the sun. Behind Aldrin is the lunar module, *Eagle*. Aldrin also took core samples, set up a seismic detector, and deployed a laser reflector. Commander Neil A. Armstrong took documentary photographs, including stereo close-ups of the surface.

339. Apollo 14. The American Apollo 14 mission made the third manned landing on the moon, in February 1971. With this mission the Apollo program shifted its emphasis from testing systems and equipment to collecting scientific data and samples. In the photograph, astronaut Alan B. Shepard, Jr., holds a double core tube sample; on the mobile equipment carrier are sampling equipment and sample bags. The geology of the collecting region, atop the Fra Mauro formation, was molded by the explosive event that created Mare Imbrium. The astronauts brought back sixty-five rocks totaling 43 kilograms (nearly 100 pounds).

340. Apollo 17. The last Apollo moon flight was made in December 1972. The seventy-two hours on the lunar surface, including twenty-two hours of extravehicular activity, gave an unparalleled opportunity for geological exploration. Geologist-astronaut Harrison Schmitt, shown here, helped obtain ninety-seven major rock samples, seventy-five soil samples, and 2,200 documentary photographs. The Taurus-Littrow landing site was chosen in order to search for both older and younger rocks than the previous missions had collected, and the team was rewarded with a richly diverse locale.

341. Skylab. Launched into orbit in May 1973, this U.S. space station was manned by three different crews in 1973 and early 1974. Skylab provided facilities for numerous astronomical and geophysical observations as well as biological experiments. Substantial data were obtained on the astronauts' physiological responses to weightlessness and on their ability to work in such an environment for prolonged periods. We see a solar panel only on the right side of the workshop area because the left solar panel was lost when Skylab was launched. Changes in the density of the earth's upper atmosphere, linked with the solar activity cycle, gradually lowered the station's trajectory, and before it could be boosted into a higher orbit, Skylab reentered the earth's atmosphere and disintegrated, in July 1979.

21

*T*he *E*xploration
*of S*pace

Mercifully—for life to exist as we know it—the earth's atmosphere blocks all forms of energy from space except visible radiation, small parts of the infrared, and much of the radio spectrum. But for the scientist this can be a problem. To observe the whole of the universe, instruments have to be carried beyond the atmosphere. Thus, rockets became vehicles in the scientist's tool kit. They are needed for direct exploration of the outer reaches of the earth's atmosphere and the near-space environment. They are also required for sending instruments to the moon and other planets.

For these reasons, the ability to do science in space owes as much to the legacy of Tsiolkovsky, Oberth, Goddard, and von Braun as does the technical achievement of men walking on the moon. The first rocket used for organized and sustained scientific studies of the upper atmosphere and near space was the German V-2 missile. Several were captured by the U.S. and Soviet armies at the end of World War II. In the United States the rockets were tested at White Sands, New Mexico. The V-2 could take a ton of scientific instruments over 160 kilometers (100 miles) into space. Previously, manned balloons carried heavy scientific devices to not much more than 20 kilometers, and unmanned balloons lofted miniature automata upward roughly 25 to 40 kilometers. The V-2, in the words of one pioneer space scientist, was an incredible bonanza.

But the V-2 was an unguided missile, not designed for scientific work. It would twist and turn

in its ballistic trajectory and spent only a few moments in space. Also, it was expensive to maintain and fly, and there were not many of them. American scientists decided to build their own scientific sounding rockets. One group, headed by James Van Allen at the Applied Physics Laboratory of Johns Hopkins University, designed the Aerobee. But in the period after the war, even the Aerobee was (by university standards) costly.

Van Allen moved to the University of Iowa in 1951, where he designed an even cheaper device. His "rockoon" was, simply, a tiny balloon-borne rocket that could carry instruments aloft to measure the distribution and intensity of charged particles at the edge of the atmosphere. Its low cost matched his meager university budget (supplemented by grants from the Navy), and the fact that it could be launched from shipboard made the rockoon an ideal vehicle because Van Allen wanted to obtain observations over a wide range of geomagnetic latitudes.

Van Allen was one of many scientists who were the motivating force behind the International Geophysical Year. While he flew his rockoons, he also prepared tiny geiger counters that he hoped might be carried aboard Vanguard, the planned first U.S. satellite, to examine globally over a long period the far reaches of the earth's charged particle environment. But the fledgling Vanguard was not ready when Sputnik flew. Van Allen's instrument was sufficiently simple, however, to be easily modified for almost any berth. Because of his experience and excellent track record as a gifted experimentalist, as well as his extensive contacts within both the Army and Navy, his experiment was the first to fly aboard Explorer, which was lofted by an Army ballistic missile called the Jupiter C. After data from Explorer 1 and 3 had been retrieved and analyzed, the Iowa group realized that they had made a fundamental discovery: the earth had vast belts of trapped charged particles surrounding it.

The first celestial object other than earth to be studied from rockets and satellites was the sun. Since it is by far the brightest star in the sky, even small instruments on spinning and tumbling rockets could measure its far-ultraviolet spectrum when they were above the atmosphere. The V-2 quickened the race to penetrate the obscuring atmospheric barrier of the earth and record the sun's high-energy spectrum. By 1947, V-2 spectrographs reached wavelengths of 2,100 angstroms in the ultraviolet, revealing for the first time solar spectra beyond the ozone barrier. Stabilized pointing controls and efficient grazing-incidence spectrometers finally reached the Lyman alpha line in 1953; as the fifties passed, photographic and electronic detectors penetrated deeper into the ultraviolet and X-ray portions of the solar spectrum, retrieving not only spectra but also full-disk images of the sun's chromosphere as seen in the light of extreme-ultraviolet emission lines.

The Orbiting Solar Observatory (OSO) series of U.S. satellites measured and monitored high-energy solar radiation. The first was launched in March 1962. Most of the instruments were developed on sounding rockets such as the Aerobee. Later satellites in the series, with improved stability, imaged portions of the solar disk, chromosphere, and corona.

The world's first manned solar observatory in space was Skylab. An afterthought from the Apollo program, it realized a dream of NASA insiders in the OSO program. Launched in 1973 by a huge Saturn booster, whose upper stage was turned into an orbital workshop and living quarters, it was manned by three successive trios of astronauts carried to the orbiting laboratory in Apollo capsules. Skylab's array of observatory-class solar telescopes, called the Apollo Telescope Mount or ATM, resurrected an old canceled, unmanned NASA satellite proposal called the Advanced OSO. ATM, powered by its paddle-wheel solar panels, monitored and imaged the sun's chromosphere and corona with ultrahigh sensitivity and resolution. Skylab images taken in 1973 and 1974 remained for years unsurpassed.

The sun is very large, even as seen from earth. Because it is so bright, its image can be magnified and examined in extraordinary detail, both from earth and from space vehicles. Other objects can be examined in similar detail only by putting large telescopes in space or by sending interplanetary probes to pay them a visit. A probe can explore directly the magnetic field and charged particles surrounding a planet. Beginning in the 1960s U.S., Soviet, and European probes with names like Pioneer, Venera, Mariner, Voyager, Vega, and Giotto visited places like Venus, Mars, Mercury, Jupiter, Saturn, Uranus, and Halley's Comet. Some of the greatest surprises came not from the planets themselves but from their moons, seen for the first time as real worlds. They were no longer faraway points of light. The rings of Saturn proved to be so complex they bordered on the mysterious; their close-up study by the U.S. Voyager probes at the beginning of the 1980s led to a new appreciation of resonance phenomena in the solar system. The features seen in the Jovian and Saturnian atmospheres and in the vast Jovian magnetosphere require continuous and close monitoring. And the brief glimpses of their satellites obtained by the Pioneer and Voyager missions justify a longer and more detailed look.

There is much left to do before we can say that we have truly explored the solar system. Planetary probes must be followed by instrumented planetary orbiters and landers. How we do this, and when, will reveal as much about ourselves as it will about the celestial neighborhood we call our solar system.

David DeVorkin

342. Van Allen belts. Something jammed the radiation counters on Explorer 1, the first U.S. artificial earth satellite. Physicist James A. Van Allen quickly redesigned the equipment and discovered that the earth is surrounded by two doughnut-shaped zones of charged particles. These particles, of extraordinarily high energy, are trapped by the earth's magnetic field. Van Allen's diagram shows the sunlit half the earth with the outlying zones of increasing radiation intensity. The hatched areas are the Van Allen radiation belts. The curved arrows mark the outbound and inbound paths of the space probe Pioneer 3.

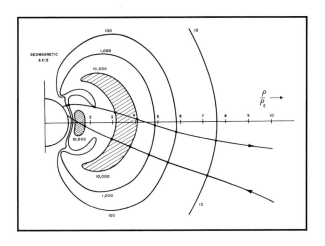

343. Van Allen. While the Soviet Sputnik caught most American scientists off guard, Van Allen had the know-how and ready equipment to mount an instrument in the first U.S. satellite, which contributed to the discovery of the Van Allen radiation belts shown above. In this mid-1957 photograph Van Allen holds one of the many Loki rockets that were launched from balloons during the International Geophysical Year to measure cosmic rays and the earth's magnetic field at high altitudes.

344. Instruments fallen back to earth. The spent spectrographic instrument package brought back ultraviolet spectra of the sun, recording a wavelength region invisible from the earth's surface. A rocket at the White Sands Missile Range in New Mexico lifted the equipment above the absorbing atmosphere for about a minute. The spectrograph and its precious data were then parachuted back, not without damage, as this 1962 photograph reveals.

345. Orbiting Solar Observatory. Beginning in 1962 the United States launched a series of OSOs into orbit around the earth. Each of these satellites, of ever increasing sophistication, carried equipment for eight to ten experiments. The OSO's horizontal base spun several times a minute to give the satellite stability. The vertical section was covered with solar cells for power and contained solar telescopes (in the long, narrow boxes) pointed at the sun. Using ultraviolet spectral lines emitted by highly energetic atoms, the OSOs explored the outermost layers of the sun's atmosphere, providing convincing evidence for the existence of "holes" in the solar corona.

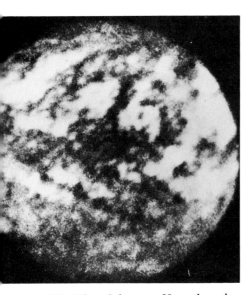

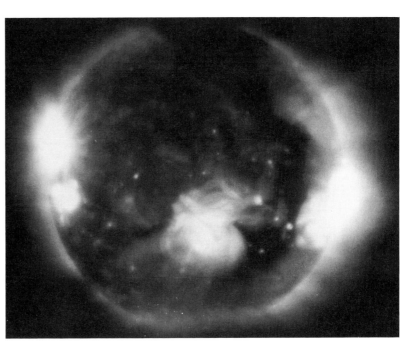

346. Ultraviolet sun. Here the solar surface is revealed in the light of the strongest hydrogen emission, known as Lyman alpha radiation. The photograph, taken in March 1959 from a U.S. Naval Research Laboratory rocket at an altitude of 200 kilometers (more than 120 miles), is the first detailed image in the extreme ultraviolet region of the spectrum.

347. X-ray sun. The X rays produced by the sun, like the ultraviolet light, are blocked out by the earth's atmosphere. Thus even crude photographs of the sun at X-ray wavelengths were not obtained until 1960, when a rocket carried a camera aloft. The remarkable picture reproduced here was taken from Skylab on June 30, 1973, when a solar eclipse was visible over Africa; pictures like this provided a valuable supplement to ground-based observations of the coronal structure.

348. A giant solar loop. This picture of the coronal region of the sun was also taken from Skylab, on June 10, 1973, a few weeks before the X-ray photograph. Enormous loops of ionized gases trapped in the sun's magnetic field had occasionally been photographed earlier, especially by ground-based coronagraphs (telescopes that create an artificial total solar eclipse for observation of the corona and prominences). Skylab carried aloft such an instrument and found that the huge gaseous loops are not so rare after all.

349. Venera 1. Beginning in 1961, the Soviet Union sent a series of probes toward the planet Venus. The first of these probes, Venera 1, went into orbit around the sun after passing by Venus at about 100,000 kilometers (60,000 miles). In 1966 a descent capsule from Venera 3 crash-landed, becoming the first spacecraft to strike the surface of another planet.

350. Ultraviolet Venus. Some probes in the American Mariner series also explored Venus. The picture shown here is a mosaic of digitally enhanced television images taken by Mariner 10 on February 6, 1974, at a distance of about 720,000 kilometers (nearly 500,000 miles) from the planet. By using ultraviolet light, the Mariner 10 cameras were able to detect cloud patterns previously undiscerned in the generally featureless disk. The swirls of cloud circulate around the planet in less than five days.

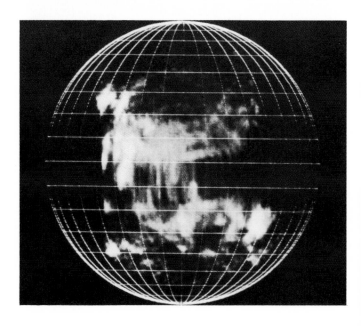

351. Radar Venus. Because of Venus's dense cloud cover astronomers were not able to discern surface characteristics until the early 1960s, when it became possible to probe the planet by bouncing radar signals off its surface. This 1969 map from the Lincoln Laboratory of the Massachusetts Institute of Technology, made with a pair of radio telescopes over 1 kilometer (about ¾ mile) apart, shows rough areas on the planet—if Venus were smooth, only a bright central spot would reflect back the 3.8-centimeter radar waves. Scientists soon correlated the signals with highlands on Venus.

ЕНЕРА-9 22.10.1975   ОБРАБОТКА ИППИ АН СССР 28.2.

352. Venus's surface. In 1970 the Soviet Venera 7 became the first spacecraft to send back signals from the surface of another planet, recording an atmospheric pressure 90 times that of the earth. The first actual pictures of the Cytherean landscape were produced by the Venera 9 and Venera 10 probes in late 1975. The image of the rock-strewn environment reproduced here was recorded by Venera 9 on October 22, 1975, minutes before the television camera perished in the 430° Celsius heat (hotter than your oven broiler!).

353. Mercury, from Mariner 10. The first close-up pictures of Mercury's surface were obtained by the American space probe Mariner 10 in March 1974. Shown here is a heavily cratered region of intercrater plains, typical of nearly half of the area photographed; the prominent foreground crater with a central peak is Glück. The scarp traversing the left center in the distance is named Victoria Rupes. About a mile, or nearly 2 kilometers, high, it is typical of a distinctive Mercurian feature not found on the moon. Probably the scarps are surface ruptures caused by the shrinking of the planet.

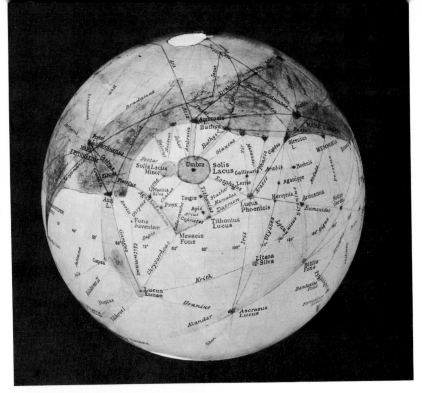

354. Lowell's Mars. Some observers in the nineteenth and early twentieth centuries, among them the American astronomer Percival Lowell, saw channels, or "canals," on Mars. The Mariner and Viking probes proved all of them to be illusory with one exception: Agathodaemon, the long diagonal line just left of center on this 1905 globe drawn by Lowell, corresponds to Valles Marineris, the giant canyon shown on the facing page.

355. Mars: the north pole. The first scientifically rewarding missions to Mars were some of the probes in the American Mariner series. This view of Mars, centered on the north polar cap, is a photomosaic constructed from 1,500 pictures taken by Mariner 9 in 1971 and 1972. The concentric ridges around the pole are the edges of permanently frozen layers of water ice; in winter the polar caps grow larger and become more prominent as dry ice—frozen carbon dioxide—condenses from the atmosphere.

356. Mars's great canyon system. The American space probes Viking 1
and Viking 2, launched in 1975, went into orbit about Mars in 1976. The
spacecraft released landers that relayed extensive data from the planet's
surface. Meanwhile, the Viking orbiters produced views of the Martian terrain from above. This detailed mosaic was constructed by the U.S. Geological Survey from 102 separate images, photographed from a range of about
32,000 kilometers (20,000 miles). Stretching 4,000 kilometers (2,500 miles)
across the dusty planet is the Valles Marineris canyon system, which vastly
dwarfs the earth's Grand Canyon in both length and depth. To the west
are three volcanoes, which, except for their enormous size (27 kilometers,
or 17 miles, high) and immense calderas, are similar to terrestrial volcanoes
like Mauna Kea in Hawaii.

357. Halley's Comet up close. In 1986, Halley's Comet made its second passage of the century around the sun. Several space probes were sent toward the comet, the European Space Agency's Giotto coming closest, to within 600 kilometers (370 miles) of the nucleus. Issuing from the dark nucleus in the upper left are two bright jets of dust. The findings from Giotto's close-up observation generally confirmed the picture set forth by the American astronomer Fred Whipple at mid-century that cometary nuclei were like "dirty snowballs."

358. Phobos. The two moons that revolve around Mars are so small that their gravity is insufficient to pull them into regular spheres. The Mariner 9 mission showed them as jagged, heavily cratered, football-shaped rocks. Phobos is the larger (about 27 kilometers, or 17 miles, long) and the closer to the planet. The Viking 2 orbiter obtained particularly good resolution in 1976, revealing strange parallel grooves or striations. The smallest detail visible is 40 meters (130 feet), across. If, as some astronomers believe, the Martian moons are actually captured asteroids, then this picture represents the first close-up of an asteroid.

359. Io. Slightly larger than the earth's moon, the innermost of the major Jovian satellites turned out to be one of the most dynamic small bodies in the solar system when probed by Voyager in 1979. Some of Io's volcanic craters, seen here, were active when the American spacecraft flew by. Forms of sulfur give the surface red and orange tints, which led to the satellite's semifacetious name of "pizza in the sky." This computer-generated mosaic was made from four sets of images taken by Voyager 1.

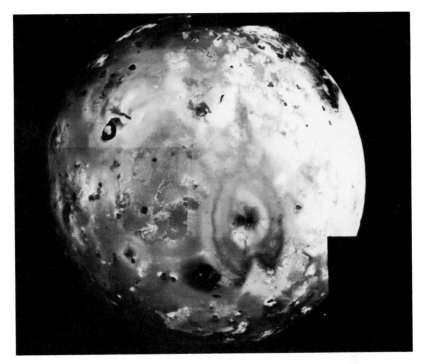

360. Mimas. Here is the innermost of Saturn's nine classical satellites, in a view produced by Voyager 1 in 1980. Heavily cratered by meteoritic impacts, Mimas has one extraordinarily large crater (*upper right*), called Herschel. Its ringed wall has a diameter of 130 kilometers (80 miles), the rims are 5 kilometers high, and the central peak 6 kilometers. The diameter of Mimas is only about 390 kilometers.

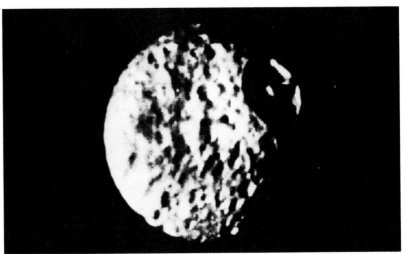

361. Miranda. This computer-assembled mosaic of images resulted from Voyager 2's passage by Uranus on January 24, 1986. The innermost and smallest of the five major moons of the planet, Miranda revealed a bizarre variety of surface features unique in the solar system. Along with old, heavily cratered rolling terrain there is young, complex terrain with sets of bright and dark bands, scarps and ridges. These young areas include the elliptical pattern at right, nicknamed "the racetrack" (or Circus Maximus), and the bright V, dubbed "the chevron."

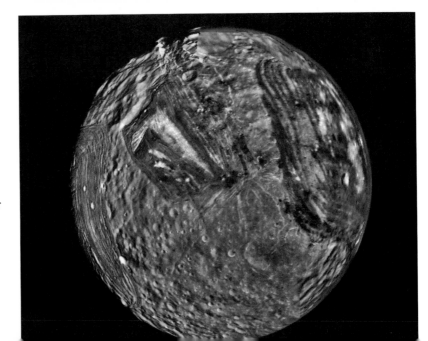

362. Jupiter's Great Red Spot. Known since the early days of telescopic astronomy, the gigantic whirlpool in the cloudy atmosphere of Jupiter has continued to puzzle theoreticians. Among the striking images produced by U.S. space exploration in the 1970s was this mosaic of the Great Red Spot, assembled from pictures taken by Voyager 1 in March 1979. The probes also produced a time-lapse film showing the counterclockwise circulation of the Great Red Spot as the entire vortex rotated around Jupiter every ten hours. Presumably the spot is an atmospheric disturbance or storm of very long duration.

363. Saturn's rings. One of the discoveries of the U.S. Pioneer and Voyager space probes was that Saturn's rings have a much more intricate structure than previously thought. Instead of just a few rings, Saturn has thousands of ringlets. This view of the so-called B ring, the second of the three classical rings, was produced from images made by Voyager 2 in August 1981. The photograph covers a range of about 6,000 kilometers (3,700 miles); the narrowest feature visible is about 15 kilometers (10 miles) wide.

364. Saturn's kinky F ring. The thin outer F ring of Saturn was discovered by Pioneer 11 in 1979. The following year, Voyager 1 showed that the ring has a complex braided structure, as we see here. The bright filaments are not more than about 15 kilometers across. Moreover, two small "sheep dog" satellites were discovered: one inside the ring, one outside. Scientists calculated that the "gravitational focusing" of these satellites could keep the particles of the ring confined and might account for the intertwining effect.

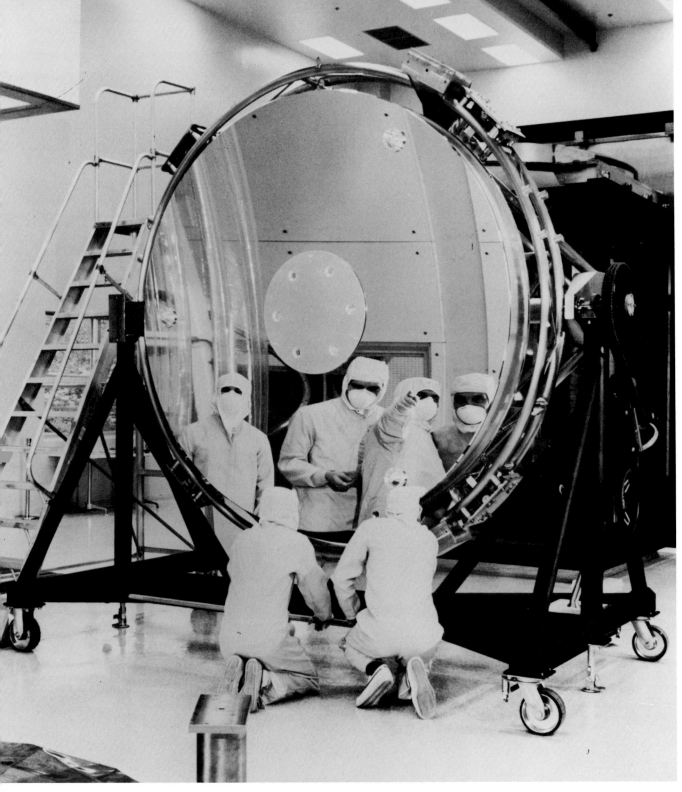

365. Hubble Space Telescope mirror. Primarily a U.S. venture and named for the American astronomer Edwin Hubble, the Space Telescope was originally scheduled to be carried aloft by the U.S. space shuttle in the mid-1980s. A telescope orbiting in space promised freedom from the blurring effects of atmospheric turbulence and access to ultraviolet and infrared light normally blocked by the atmosphere. Here we see the 2.4-meter (94-inch) primary mirror, probably the most precise large optical surface ever made.

CHAPTER

22

The Depths
of the Universe

In this and many other ways great modern instruments have rapidly advanced our knowledge of the structure of the universe and enabled us to sound its depths and to trace the evolution of the stars.

George Ellery Hale wrote those words in a 1922 essay entitled "The Depths of the Universe." What was true then is even more relevant as the twentieth century draws to its close. Hale mentioned the 100-inch and 60-inch reflectors on Mount Wilson, and the depths of the universe to which he referred were at most 10 million light-years away. Today still larger telescopes of assorted kinds have plumbed the depths a thousand times farther.

Hale did not live to see his largest project completed. The Palomar 200-inch, delayed by World War II, did not go into operation until 1949. It could probe deeply, but not widely—estimates suggested that it would take centuries to map the northern sky completely. Responding to pressures for more available telescope time, astronomers at the Lick Observatory created a 120-inch (3-meter) telescope, completed in 1959, which used one of the glass blanks left over from the Palomar project. Around the same time, a group of American university astronomy departments formed a consortium to promote a national optical observatory, which was established on Kitt Peak southwest of Tucson, Arizona. Eventually the Kitt Peak National Observatory built 4-meter-class reflectors in both the northern and southern hemispheres: on Kitt Peak, in 1973, and at Cerro Tololo in Chile, in 1976.

Western European astronomers, facing decidedly poorer meteorological conditions, nevertheless found sites in their countries for intermediate-sized instruments. As a point of national pride, British astronomers set up the 98-inch (2½-meter) Isaac Newton telescope under less than optimum skies in Sussex. When they subsequently had an opportunity to lead the world in radio astronomy with a huge steerable antenna at Jodrell Bank near Manchester, their devotion turned to the radio wavelengths; eventually they transported their 98-inch reflector to the palmier skies of the Canary Islands, where by the late 1980s it was joined by the 4-meter William Herschel reflector. In the meantime, the British collaborated with the Australians to build a 3.9-meter telescope at Siding Spring in New South Wales. Other European astronomers, as well, saw an open frontier in the southern skies, and they banded together to found the European Southern Observatory with an observing site at Cerro La Silla in Chile. By 1976 they had opened a 3.6-meter telescope there. Yet another international effort brought together the Canadians, French, and Americans, who placed a 3.6-meter instrument on Mauna Kea in Hawaii; it saw first light in 1979.

The biggest single-element optical telescope is the Soviet 6-meter (236-inch) reflector in the Caucusus Mountains in southern Russia. Hampered by problems related to its size, as well as those of limited auxiliary instrumentation, the giant eye got off to a slow start after its 1975 opening. Nevertheless, it successfully pioneered the idea of an altitude-azimuth mounting, as opposed to the equatorial arrangements previously favored, in which one axis of rotation was aligned parallel to the earth's axis.

The effectiveness of these huge telescopes was enormously enhanced during the 1970s and 1980s by the development of electronic devices that captured a far greater proportion of the scarce cosmic photons than the photographic plates that were long the workhorse detector for astronomical data gathering. Thus a comparatively modest reflector, such as the 61-inch at Harvard, Massachusetts, became a more effective light-collector than the 200-inch had been in 1950. Armed with such tools, astronomers with the largest telescopes have been able to probe deeper and deeper. If in the 1970s many astronomers supposed that 2 billion light-years might represent a practical limit for the detection of ordinary galaxies, by the end of the 1980s the record exceeded 10 billion light-years.

The quest for ever-greater penetration into the recesses of our Milky Way galaxy, as well as into the deep space beyond, was not limited to the wavelengths of visible light. Three of the most fascinating discoveries of the 1960s depended on radio astronomy: the pulsars, found by radio astronomers working in Cambridge, England; the quasars, first identified by optical astronomers who were examining radio sources; and the microwave background radiation. The pulsars turned out to be rapidly spinning neutron cores of exploded supernovae. Detected within our own galaxy, they bore witness to the explosive fireworks of past ages, when massive, rapidly evolving stars grew old and died in violent paroxysms. The quasars, the collapsing nuclei of distant galaxies (eventually to become immense black holes, in the opinion of many theoreticians), were the most energetic phenomena known. And the background radiation was the measurable remnant of the primordial explosion that gave birth to our universe. It is no wonder that astronomers of the 1970s began to speak of "the violent universe."

Reinforcing this view were the findings of the high-energy astronomers, who concentrated on X rays and gamma rays, at the high-energy end of the spectrum. Satellites sensitive to this energetic radiation flew above the protecting shield of the earth's atmosphere to detect unexpected sources of celestial X radiation. One of their most interesting findings was a series of candidate objects that might betray the presence of black holes.

In the 1970s particle physicists teamed up with astronomers to describe the Big Bang in its opening moments. In calculating backward in time to the moment when the expanding universe was compressed into an unimaginably dense, energetic ball, they found temperatures so high that atoms could not survive. A mere 10^{-23} second from time zero, the presently known laws of physics no longer applied. At that point there was pure electromagnetic energy, which began to form matter according to Einstein's famous $E = mc^2$ equivalence of energy and mass. Based on this, scientists made a prediction as to the proportion of helium in the universe, obtaining a number verified by repeated observations. The scenario also incorporated the idea that eventually—about 300,000 years after the Big Bang—the universe would become transparent and that photons of light from the initial fireball should still be flying through space. These relic photons account for the microwave background radiation first observed in 1965.

Still puzzling to astronomers was the process of galaxy formation, something that took place between the primordial fireball and the epoch of the quasars, perhaps 10 billion years ago. Studies of the spatial arrangements of galaxies, which became a prime target of research in the 1980s, helped gain insight into the spongelike distribution of these massive aggregations of stars, dust, and gas. But precisely why the galaxies were formed in such patterns, with huge voids, remained elusive. The search for answers motivated astronomers to plan the Hubble Space Telescope, a 2.4-meter (94-inch) orbiting mirror that would function above the earth's murky, tremulous atmosphere. Meanwhile, around the world astronomers began planning a new generation of still larger ground-based telescopes. In 1979 a pioneering high-technology telescope, a multiple-mirror instrument, went into operation on Mount Hopkins, south of Tucson, Arizona. Following this lead, new telescopes would employ mosaic or multiple-mirror systems to allow still more light-gathering power as astronomers pursued the elusive photons that might reveal the past and, by implication, the future of the universe.

Owen Gingerich

366. 120-inch Lick reflector. When completed in 1959, this large telescope at the Lick Observatory on Mount Hamilton was exceeded only by the 200-inch on Mount Palomar. Younger, far-sighted members of the Lick staff had pushed for the largest feasible telescope that could be funded by the California Legislature, and they cut costs by using the 120-inch (3-meter) glass test blank left over from the Palomar project.

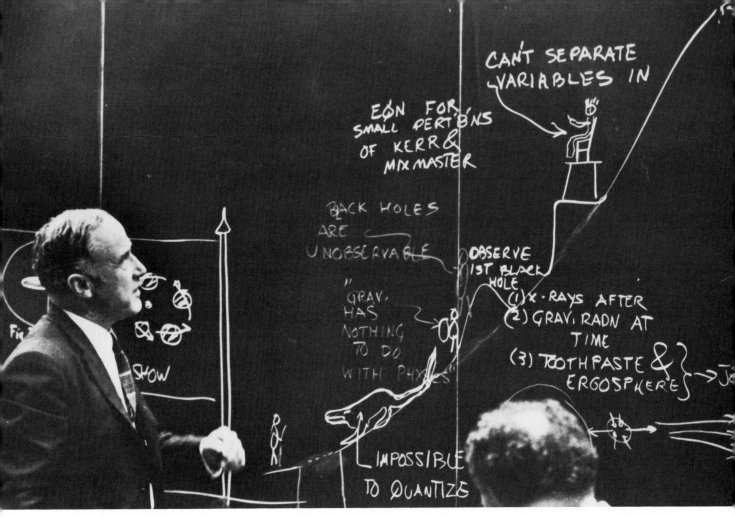

367. Wheeler lecturing on black holes. Einstein's general theory of relativity allowed the possibility of regions of space containing matter so condensed that the escape velocity was greater than the speed of light. These were named black holes by the American physicist John Wheeler, who had moved from work on fission and thermonuclear weapons to cosmological research. In the classical view, nothing can escape from a black hole, so powerful is its gravitational attraction. Necessarily, the structure of a black hole must be extraordinarily simple: "A black hole has no hair," said Wheeler.

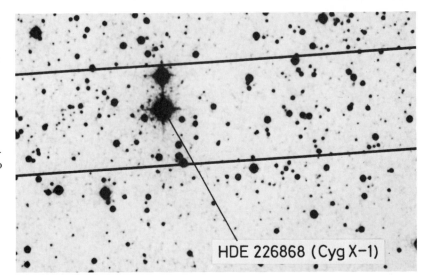

HDE 226868 (Cyg X-1)

368. Cygnus X-1. How to detect a black hole? Astronomers reasoned that if a black hole happened to form a binary system with a star, certain phenomena could be observed. The black hole, for example, might capture gas from its companion, and the rapidly accelerated gas would produce intense X rays. Beginning in the 1970s astronomers discovered in binary systems several X-ray sources that might be black holes. The first extremely promising candidate was Cygnus X-1, which was found to be a massive companion of the bright star HDE 226868.

369. Bell and the Four-Acre Array.
A totally unexpected phenomenon
was discovered in 1967 by Jocelyn
Bell, then a graduate student working with Antony Hewish at this
Cambridge University radio telescope. She noticed recurrent radio
signals, sharp pulses at precise intervals, coming from a certain direction in the sky—the first pulsar
was found. Several hundred pulsars, apparently the rapidly rotating
neutron cores of bygone supernovae, were subsequently discovered
by astronomers.

370. Supernova 1987A. A violently
exploding star in the Large Magellanic Cloud flared into view in February 1987, giving astronomers
their closest look at a supernova in
nearly four hundred years. The
pair of before and after pictures
below show the 30 Doradus Nebula
region of this satellite galaxy to our
Milky Way; the second was taken
soon after the discovery. The event
provided an opportunity to confirm
theoretical predictions about the
evolution of such profligate exploding stars.

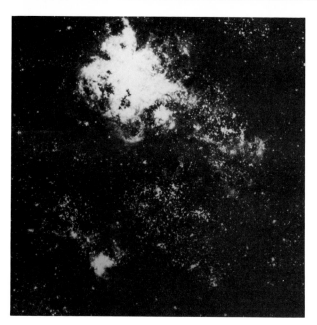

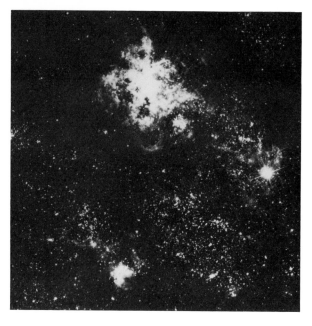

371. Kitt Peak. Established in 1958 with support from the National Science Foundation, the Kitt Peak National Observatory in Arizona provides large-scale instrumentation for astronomers whose institutions cannot afford major optical telescopes. In this time-exposure photograph, taken during a spectacular thunderstorm, a fork of lightning frames the nineteen-story dome of the Mayall telescope. When completed in 1973, that 4-meter (158-inch) reflector was the world's third largest telescope. To the right, in front of the Mayall reflector's dome, is the one for a 50-inch reflector. A major solar telescope is also sited on Kitt Peak.

372. Cerro Tololo 4-meter reflector. The extraordinarily favorable southern skies over Chile's Andean desert make it an enticing setting for astronomers. The major astronomical facilities sited there include the Cerro Tololo Inter-American Observatory, established in 1965 as a branch of the Kitt Peak National Observatory. Cerro Tololo's 4-meter reflector began observing the southern skies in 1976, three years after its northern twin had joined the astronomical arsenal.

373. European Southern Observatory. Another major center for study of the southern skies is the ESO facility at Cerro La Silla, near Cerro Tololo. Supported by several European countries, the ESO facility was opened in 1965. A 3.6-meter reflector went into operation in 1976. In 1987 the ESO undertook to construct an array of four 8-meter mirrors, which, working in tandem, can give a light-gathering power equivalent to a single 16-meter telescope.

374. Stepping stone into deep space. This photograph of a section of the giant cluster of galaxies in the constellation Virgo was taken with Kitt Peak's 4-meter reflector. At a distance of roughly 50 million light-years, the Virgo cluster is the closest large grouping of galaxies and the hub for our region of the universe. Nevertheless, it lies so far that distance indicators such as Cepheid variable stars had as of late in the century still escaped detection.

375. Hercules cluster. The distant spiral and elliptical galaxies of this remote cluster in the constellation Hercules are scattered among the faint foreground stars. Also photographed from Kitt Peak with the 4-meter reflector, in 1973, these star systems are about ten times as far away as the Virgo cluster.

376. Galaxy cluster in Hydra. At 3 billion light-years, these faint galaxies in the constellation Hydra appear as small and indistinct fuzzy spots concentrated toward the center of the picture. They are almost lost among the foreground stars of our own Milky Way galaxy. The Hydra cluster was one of the most distant photographed in the early days of the Kitt Peak 4-meter reflector. At a redshift of 0.2, its galaxies are rushing away at 60,000 kilometers (nearly 40,000 miles) per second.

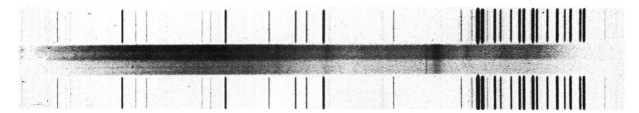

377. Colossal speeds in deep space. In 1963, while studying the baffling spectrum of the quasar known as 3C 273, Caltech astronomer Maarten Schmidt realized that its peculiar features were hydrogen lines with an unprecedented redshift, indicating a remote object receding at high speed. Above, red to the right, is Schmidt's fuzzy spectrum for 3C 273 (taken two days after the overexposed original spectrum), flanked by sharp-lined comparison spectra.

378. Most distant object. On this photographic negative, taken in 1974 with the U.K. Schmidt telescope in Australia, the arrow marks the position of the quasar that, when it was identified in the early 1980s, was the most distant observed object. The quasar, PKS 2000-330, exhibited an extraordinary redshift of 3.78, compared with about 1 for the most distant galaxies then known. Later in the 1980s quasars with redshifts greater than 4 were found.

379. Multiple Mirror Telescope. The MMT was the first of a new generation of ground-based telescopes designed to increase light-gathering power while avoiding the high costs and physical difficulties posed by huge glass disks. It links six 72-inch (1.8-meter) mirrors to create a single image equivalent in intensity to that produced by a single 176-inch (4½-meter) telescope. Dedicated in 1979, the MMT sits atop Mount Hopkins, south of Tucson, Arizona.

380. Mauna Kea. The dormant volcano Mauna Kea on the island of Hawaii is the location of several large reflectors of the 4-meter class, to be joined in the 1990s by the 10-meter mosaic-mirrored Keck telescope (at a site just off the center bottom of this air view). Hawaii's southerly location and the high, dry skies available from the 4,200-meter summit (13,800 feet) have made the mountain a mecca for both American and European astronomers.

381. Soviet 6-meter (236-inch) reflector. When Soviet astronomers embarked on this ambitious project in 1960, they pioneered in the modern use of the altitude-azimuth mounting rather than the more familiar equatorial form, where one axis points to the celestial pole. The 70-ton mirror, cast of specially developed low-expansion glass, was ground and polished at the Leningrad Optical Works beginning in 1968. Seven years later, from its location in the Caucusus Mountains in southern Russia, the giant camera took its first sky photographs.

382. The 6-meter mirror cell. The enormous size of the world's largest single telescope mirror is clearly felt here. The circular break in the floor delimits the revolving azimuthal platform, on which the entire instrument turns. Hundreds of workers maintain this operation, all part of Big Science.

383. Twin Tail radio galaxy. The radiograph of a galaxy that is a strong emitter of radio waves may reveal immense lobes or jets extending far beyond its nucleus. A case in point is this image of radio source 3C 75, made with the Very Large Array in New Mexico. The source, associated with the twin nuclei of the central galaxy in the cluster Abell 400, has a pair of giant jets, possibly wrapped around each other. The jets are bent, perhaps from a vacuous wind of hot gas within the galaxy cluster.

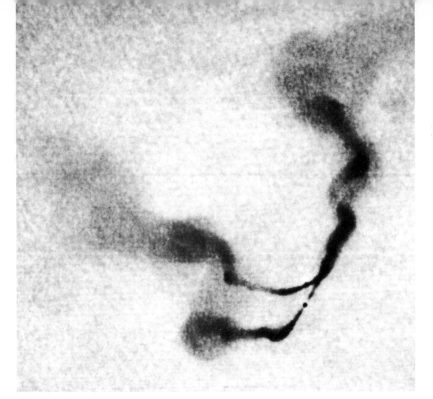

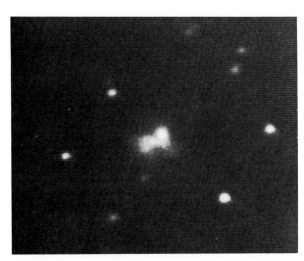

384. Cygnus A: Colliding galaxies? First detected around 1946, the most powerful source of radio waves in the constellation Cygnus was also the first radio source found to lie outside the Milky Way, at a distance of half a billion light-years. When precise positions were established for Cygnus A, it was linked to the curious optical object at left. Astronomers at first suspected that it was a pair of galaxies in collision but then came to see it as a single elliptical galaxy cut by a dark dust lane.

385. Radio view of Cygnus A. The spectacular and detailed radiograph below is from the Very Large Array. Between the two radio lobes lies the giant elliptical galaxy, not visible here. That galaxy is centered at the dot at the center, where the energy that powers the radio source appears to originate. Thin jets of fast-moving gas connect the nucleus with the lobes.

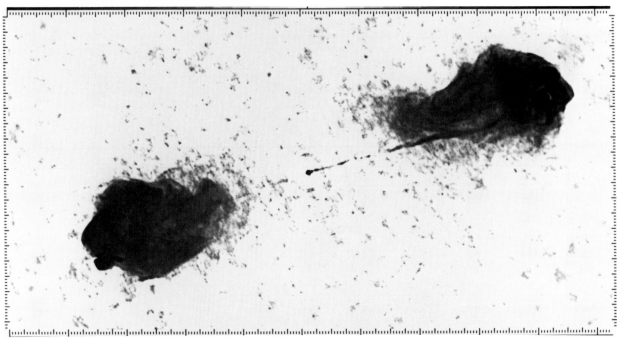

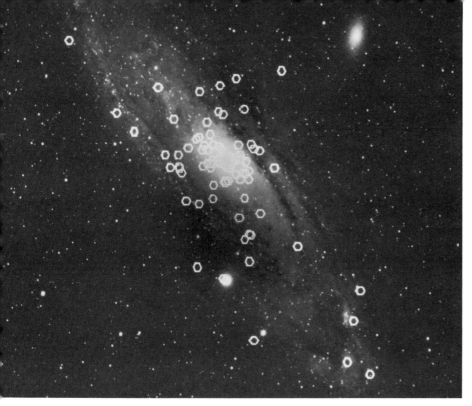

386. X-ray sources in the Androm-eda galaxy (M31). The Einstein high-energy observatory satellite, launched by the United States in 1978, made it possible to obtain images of the heavens in X-ray light. The circles on this optical image of the galaxy's central regions indicate some of the 72 X-ray sources found in M31.

387. A whisper from space. In 1965, Arno Penzias and Robert Wilson, working with this horn antenna at the Bell Telephone Laboratories in Holmdel, New Jersey, detected a uniform background static at a wavelength of 7 centimeters. The radiation appeared to be a remnant of the primeval fireball that resulted from the Big Bang, and its discovery strengthened the case for an evolutionary cosmology.

Part Eight

THE
SCIENCE
OF
SCIENCE

388. Rome Conference, 1931. Nuclear physics had come of age. The first international conference devoted to the field was held in Rome in October 1931, on the eve of the glorious year 1932, when there would come reports of the discovery of deuterium, of the neutron, of the positron. Among the eminent scientists assembled in Rome for the meeting were the three prominent in this photograph: Robert Millikan, Marie Curie, and Werner Heisenberg.

CHAPTER

23

Scientists Unite!

In the seventeenth century scientists in Italy, England, France, and Germany founded societies to foster the discussion and dissemination of research results. These organizations provided a channel for government support of science, especially for large expensive projects, and in some cases their members received salaries. The societies occasionally offered substantial prizes for papers on specified topics, thereby influencing the direction of research.

By the nineteenth century many of the established societies were no longer serving their original purpose; they were dominated by older scientists and wealthy amateurs who not only failed to do original research themselves but sometimes prevented good research from getting published. Having been founded in capital cities such as London and Paris, the societies did not offer much encouragement to scientists located in the provinces. In most cases they refused to admit women; Marie Curie was denied election to the French Academy of Sciences even after she won the Nobel Prize.

New societies appeared. These—such as the British Association for the Advancement of Science, founded in 1831—were typically open to everyone interested in science and met in a different location each year. New journals like the *Philosophical Magazine* and *Nature* (both British) and the *American Journal of Science* were more willing to accept papers by unknown scientists and published some of the most important discoveries of

the nineteenth century. More specialized societies for physics, chemistry, astronomy, and geology began to proliferate later in the century; they were needed especially in the physical sciences, because the proceedings of general scientific societies tended to be dominated by biology unless split into separate sections.

The twentieth century saw a rapid increase in the number of scientists and of papers in physics and chemistry, which stimulated a proliferation of independent specialized journals. The societies, however, gradually took on the burden of publishing the bulk of the research literature. At the same time the center of gravity of scientific activity shifted from Europe to the United States. Early in the century the most important physics papers appeared in the independent journals, such as *Annalen der Physik* and *Philosophical Magazine*, but during the 1930s *Physical Review,* sponsored by the American Physical Society, became the largest and most influential journal. Chemists at the beginning of the century relied on the *Berichte* (reports) of the German Chemical Society and the *Journal of the Chemical Society of London;* later the *Journal of the American Chemical Society* became the dominant publication. Interdisciplinary research came to be announced most often in periodicals such as the *Journal of Chemical Physics* (published by the American Institute of Physics) and the *Journal of Geophysical Research* (published by the American Geophysical Union). Astronomy did not follow the trend of increasing dominance by society publica-

259

tions. The independent *Astrophysical Journal* (founded by George Ellery Hale and James E. Keeler in the United States) was the most important periodical, while the *Monthly Notices of the Royal Astronomical Society* gradually lost the prominence it enjoyed in the preceding century. One independent general magazine, *Nature,* remained successful in attracting original reports of some of the outstanding discoveries in all areas of science, thus resisting the trends toward specialization, society publishing, and American dominance.

Early in the twentieth century scientists recognized the need for new international organizations to encourage and coordinate worldwide research projects and conferences. After World War I "international unions" were established in fields such as pure and applied physics (IUPAP), pure and applied chemistry (IUPAC), geodesy and geophysics (IUGG), and astronomy (IAU). An International Research Council (later renamed the International Council of Scientific Unions) was set up to oversee their activities.

In the years after World War II the congresses and other meetings sponsored by the international unions advanced the progress of science in two important ways that may not have been anticipated by their founders. First, they facilitated communication and joint research projects between Western and Soviet-bloc scientists, who were often kept apart by their hostile governments. Second, they provided opportunities to developing countries for work in areas that would otherwise have been dominated by the technically advanced countries.

The international unions played a major role in sciences involving the coordination and analysis of worldwide observations—sciences such as meteorology, oceanography, and terrestrial magnetism and upper-atmosphere studies. The unions' most spectacular success was the International Geophysical Year (IGY), extending from July 1, 1957, to December 31, 1958. (This period was chosen because it was the next expected peak of solar activity and would provide an opportunity to observe intense electromagnetic phenomena in the earth's atmosphere.) The IGY was conceived at a 1950 brainstorming meeting of geophysicists, brought together in the Maryland home of physicist James Van Allen to confer with a British visitor, Sydney Chapman. Chapman, who had pioneered the mathematical analysis of terrestrial magnetism and the physical properties of the upper atmosphere, was soon chosen to head the committee that administered the IGY, under the auspices of the International Council of Scientific Unions.

The Soviet Union, though not a member of the council, was invited to participate in the IGY and announced its intention to launch artificial satellites to probe the earth's ionosphere. The Soviets put their first artificial satellite, Sputnik, into orbit in October 1957, thus beginning the Space Age. The following year the United States succeeded in launching several satellites into earth orbit. They provided the first major scientific discovery of the IGY: the Van Allen radiation belts circling the earth. The Soviets topped this in 1959 by sending a rocket to photograph the backside of the moon.

In addition to inaugurating the first major phase of space exploration, the IGY led directly to the establishment of the modern theory of plate tectonics, as research begun during the IGY provided much of the stimulus for the revolution in the earth sciences that started in the 1960s.

Scientists also united, especially in the later decades, to try to persuade politicians and the public to avert threats to human survival and to the environment arising from military and commercial applications of science. Following the use of the atomic bomb against the Japanese in World War II, American scientists lobbied successfully for civilian rather than military control of atomic energy in the United States, although the result of their efforts, the Atomic Energy Commission, was later criticized for many of its actions. In the 1950s the dangers of radioactive fallout from nuclear tests in the atmosphere became apparent; the American chemist Linus Pauling in 1958 presented to the United Nations a petition signed by over 9,000 scientists urging a ban on such tests. A test-ban treaty, forbidding nuclear explosions in the atmosphere and also under water and in space, was finally concluded between the United States, the Soviet Union, and Great Britain in 1963.

Meanwhile, the American biologist Rachel Carson had the year before published a controversial book, *Silent Spring,* warning of the dangers of chemical pesticides. Galvanized into action, scientists organized campaigns to fight environmental hazards and to preserve endangered species. They argued, in effect, "The species you save might be your own!"

Stephen G. Brush

260

389. Göttingen Physical Society, 1907. The University of Göttingen became particularly strong in physics and mathematics in the late nineteenth century, presaging its wide influence in the twentieth century. Among those in the bottom row, for example, are the mathematician and physicist Carl Runge (*second from left*), physicist Woldemar Voigt (*third from left*), and physicist Max Abraham (*far right*).

390. A league of eminent scientists. Several distinguished scientists served on the International Commission on Intellectual Cooperation established by the League of Nations after World War I. Among those present in this 1927 photograph are Robert Millikan, Albert Einstein, and Hendrik Lorentz, at that time the commission's president.

391. Physicists amass. Here are all those attending the spring 1930 meeting of the American Physical Society in Washington, D.C. By the early 1930s the APS, which had been founded in 1899, was the largest physics organization in the world.

392. Masses of physicists. Such was the growth of the scientific enterprise: in just a few decades the number of practicing physicists increased enormously, as evidenced by attendance such as this, at one of several simultaneous sessions at an APS meeting in the 1970s.

393. Kazan, 1928. Among those present at this Soviet-sponsored congress of physicists were the Austrian-born mathematician Richard Von Mises, the British physicist Paul Dirac, the American Homer Dodge, the Soviet Abram Feodorovich Ioffe, the Dutch-born physical chemist Peter Debye, and the German physicist Max Born.

394. First Solvay Conference. The Belgian industrial chemist Ernest Solvay established a series of international meetings of distinguished physicists held in Brussels. The first took place in 1911, with Solvay seated at the head of the table. Others present include Einstein, Kamerlingh Onnes, and Rutherford (*standing, second, third, and fourth from the right*), Planck (*standing, second from left*), and Lorentz (*to the right of Solvay*). Marie Curie and Henri Poincaré are seated at right.

395. Rockets for high-altitude research. The International Geophysical Year, from July 1, 1957, to December 31, 1958, was the largest international scientific enterprise that had ever been undertaken. The purpose was to study all aspects of the physical environment in a period that coincided with a maximum in the eleven-year sunspot cycle. Massive amounts of data were collected during the IGY and its extension in 1959. The IGY saw the first use of rockets for high-altitude measurements during an eclipse: here are Nike-Asp rockets, aboard a U.S. ship in the South Pacific, preparatory to launching at the eclipse of October 12, 1958.

396. Giant balloon for stratospheric research. During the IGY, several nations, including the United States, pioneered in stratospheric meteorology with balloons that could chart the circulation at levels above 20 miles, or about 32 kilometers. This 411-foot (125-meter) Winzen balloon about to be launched from the USS *Valley Forge* was photographed in 1960. It was an example of the high-altitude probes used earlier in the IGY; the intention here was to study cosmic rays.

397. Japanese moonwatch team.
When the first artificial earth satellites were launched during the IGY, provisions for tracking them were in a rudimentary state. Project Moonwatch, organized by the Smithsonian Astrophysical Observatory, enlisted hundreds of amateurs throughout the world, who worked in teams. Each team patiently watched with fixed scopes for the elusive new satellites to cross the meridian at its location.

398. Seismic research in the Antarctic. IGY studies focused not only on the atmosphere and space but also on the oceans and the earth's structure. Seismic sounding techniques were used in the Antarctic to yield substantial data on the snow and ice cover, as well as the underlying land mass. Here we see an explosion used to measure ice thickness.

399. Mirabelle. An example of international cooperation in nuclear physics is this particle detector, used in conjunction with the large accelerator at the Soviet Institute of High-Energy Physics in Serpukhov, just south of Moscow. The first of a new generation of large liquid-hydrogen bubble chambers, with a volume of 10 cubic meters, it was built by French scientists and engineers at the nuclear research center in Saclay, France. Installed in 1971, Mirabelle produced thousands of photographs in the 70-gigaelectron volt proton beam of the Soviet facility.

400. Eisenhower and atoms for peace. Amid an intensifying nuclear arms race, U.S. President Dwight D. Eisenhower proposed to the United Nations General Assembly a plan for international cooperation in the peaceful use of atomic energy. In the years following his December 1953 address, cooperative agreements were negotiated between various nations concerning the peaceful use of the atom in industry, medicine, agriculture, and scientific research. In 1957 the International Atomic Energy Agency was formed to foster the role of atomic energy in advancing world peace and welfare.

401. Scientists protest. "Science for the people" was the watchword of these activists at the 1970 meeting of the American Association for the Advancement of Science as they sought to influence the use to which scientific research was put by society at large. Watchdog groups of scientists have continued to monitor developments in environmental protection, human rights, and biological engineering.

402. Protesting nuclear testing. Despite the benefits of the Atoms for Peace program, the nuclear arms race continued unabated. In April 1962, Linus Pauling picketed the White House with other demonstrators protesting the resumption of nuclear tests in the atmosphere by the United States. Pauling, who had received the Nobel Prize in chemistry in 1954 for his work on the chemical bond, won the Nobel Peace Prize in 1962 for his efforts toward a nuclear test ban treaty.

267

403. Proliferation of scientific literature, 1939. The twentieth century has seen the rise of Big Science, an explosion in the number of scientists, and an acceleration in the pace of research. All this has meant more papers, articles, monographs. How could the individual scientist keep up? How could editors and publishers ensure the quality of what they publish? The problem was already serious in the first decades of the century, as this cartoon from a German geological journal indicates.

404. Proliferation of scientific literature, 1966. By the second half of the century, the situation, naturally, was worse. Better technology had brought still better means of gathering data. This cartoon is from an American engineering publication.

405. The growth of a journal. The leading American journal of physics, *Physical Review*, has appeared throughout the century. Here *Physical Review* deputy editor in chief Peter Adams is sitting with copies of the journal from two years about five decades apart. At right are the issues for 1931; at left, for 1985. Such was the expansion of physics that around 1970 the journal split into separate series for different fields; one dealt with general physics, but others were devoted to specialized topics such as solid-state physics; a companion publication called *Physical Review Letters* began in 1958.

406. Vostok on exhibition. Vostok 1 carried the first man into space—Yuri Gagarin—in 1961. Scientific and technological displays, such as a Vostok spacecraft, are found in Moscow at the huge Exhibition of the Achievements of the National Economy of the U.S.S.R. A permanent exhibition, established in 1959 on the basis of several existing pavilions, it presents Soviet accomplishments in industry, construction, agriculture, culture, and public health.

Communicating Science

Scientists have always been communicators, mostly within their own community of scholars, but some to the wider world. In the early nineteenth century the illustrated lectures and demonstrations of Michael Faraday at Britain's Royal Institution thrilled and fascinated one and all with the wonders of electricity. As Faraday and others were motivated to put before the public the way the natural world works, they acquired physical apparatus that manipulated, transformed, and displayed reality as they saw it in terms of physical law. These instruments—air pumps, thermometers, barometers, batteries, spectroscopes, telescopes, microscopes, and mechanical analog devices of all forms and purposes—became wondrous objects of interest in themselves. Not only did they reveal the workings of the natural world, they demonstrated the ingenuity of the scientist and instrument maker in capturing the essence of perceived physical law. By the end of the nineteenth century these devices were the corpus of science. For the public they became icons of mankind's mastery of the physical world.

These devices, these icons, had to be preserved, displayed, and interpreted for the public. The small, enclosed arenas of classrooms could not fill this need, but museums could. Two distinct styles of science museums emerged in the twentieth century. The first type treated scientific instruments as three-dimensional documents characterizing the technology of science (the instruments of observation) and showing how that technology

changed over time. Later came the "science center," wherein concepts and principles could be deduced from hands-on experimentation, performed either by a guide or, as in San Francisco's Exploratorium, by the visitor. In addition to the instruments of science, the products of industry could be instructive objects for display. Such is the case at Moscow's huge multipavilion Exhibition of the Achievements of the National Economy of the U.S.S.R.

Simultaneous with the growth of science museums was the development of the optical projection planetarium. Traditional mechanical planetariums had freed the observer from the bounds of space and time, letting days, months, and years pass with the turn of a crank, but they placed the observer outside the solar system, looking in. This problem of perspective was partially solved by Wallace W. Atwood of the Chicago Academy of Sciences around 1910. He revived an old idea: placing people inside a hollow perforated sphere to view the stars. Atwood designed a 15-foot equatorially mounted star sphere for the academy's museum that visitors could walk into to view the night sky.

In Germany at about the same time, Oskar von Miller, the Deutsches Museum's founder and director, counseled by the astronomer Max Wolf, was also searching for an improved planetarium design. Walther Bauersfeld of the Carl Zeiss Optical Works suggested projecting images of celestial bodies on the inside surface of a stationary sphere.

Such an optical projection planetarium offered the possibility of a powerful illusion, with the terrestrial observer looking out at a familiar sky. The embodiment of this new concept came at Bauersfeld's hands in 1919; his first projector was installed in the Deutsches Museum in 1923. Within a few years, Zeiss made a second and much larger model with two projection star spheres and dual projectors for the sun, moon, and planets that would produce bright, realistic images in a dome large enough to seat hundreds of people. The critical and popular success of the devices ushered in a new era of popularizing astronomy.

Communicating astronomy to the public in planetariums began more as a celebration of the ingenuity of the Zeiss planetarium projector. By the 1950s planetarium lecturers were emphasizing increasingly sophisticated special-effects devices to provide depth to the two-dimensional visible universe the Zeiss projector presented. Travels through space as well as time became the rage. But planetariums, along with museums of science, allowed the public to witness the scientific enterprise in a manner that transcends the written word. Today, in an era when museums and planetariums are compelled to compete with extremely realistic and captivating two-dimensional images of reality and fantasy on film and television, they continue to provide a three-dimensional view: an extension of the scientific laboratory, if not of the mind of the scientist.

Combining celebration with education, fantasy with reality, circus sideshow with museum, international expositions and world's fairs have long reflected the hopes and aspirations of nations. Driven by national ambitions but couched in terms of industrial and social progress, expositions became a worldwide movement after the success of the Great Exhibition of the Works of Industry of All Nations, held at London's Crystal Palace in 1851. In the United States especially, they provided a means through which a rapidly expanding young land could define and present itself as a society of world importance.

Almost one-fifth of the population of the United States passed through the gates at the Centennial Exposition, held in Philadelphia in 1876. Visitors were shown America at work: marvelous high-speed machines that formed 80,000 screws a day represented modern society. The World's Columbian Exposition in Chicago in 1893 continued this theme, though in utopian terms. Displays of sci-

entific devices of American manufacture offered powerful testimony to the capabilities of this new land. International congresses in science and letters convened against the backdrop of Warner & Swasey's gigantic iron mounting for the largest telescope in the world, George Ellery Hale's 40-inch refractor for the Yerkes Observatory.

Telescopes gave way to airships, automobiles, ice cream cones, and anthropology at the 1904 Louisiana Purchase Exposition in St. Louis. Here the scientific congresses were even more international in scope. Savants traveled to St. Louis from around the world to review the past century and to point the way into the next. The germ of an international organization to coordinate research in astronomy was born here, through the labors of George Ellery Hale, influenced by the Dutch astronomer Jacobus Kapteyn. Images of the future became a stronger theme in later expositions and fairs. The future's promise was heavily promoted at the Century of Progress International Exposition in Chicago in 1933–1934 and the New York World's Fair of 1939–1940. Both fairs celebrated science in the cause of mankind: science, working together with corporate America, would solve the woes of the Depression and lead to a new life. Science was again at the forefront at the first major world's fair after World War II, the 1958 Brussels exposition Expo '58, held amid the International Geophysical Year. The fair's symbol, the Atomium, reflected the theme of the peaceful use of atomic energy. Among later fairs, Japan's Expo '85 at Tsukuba Science City near Tokyo featured an abundance of robots and projected images, including those on a mammoth television screen measuring 77 feet by 124 feet (about 23 meters by 38); the theme was science and technology for man at home.

World's fairs have come to offer gigantic idealized models of reality inviting spectators to view the latest technology for science and the welfare of humanity. Visitors see exhibits and demonstrations of scientific principles and practice, where system and order are revealed. The ultimate message of the fairs is that for society to advance, those parts of society responsible for such revelations should be appreciated and supported. World's fairs portray scientists and engineers as essential and central ingredients of a vital, competitive culture.

David DeVorkin

407. Engineering gallery in London Science Museum. Founded in 1857, the Science Museum at South Kensington has one of the world's most celebrated collections of objects pertaining to the history of the physical sciences and technology. Famous locomotives, turbines, telescopes, and computing machinery vie for space. This gallery, crowded full of engines, was photographed around 1910.

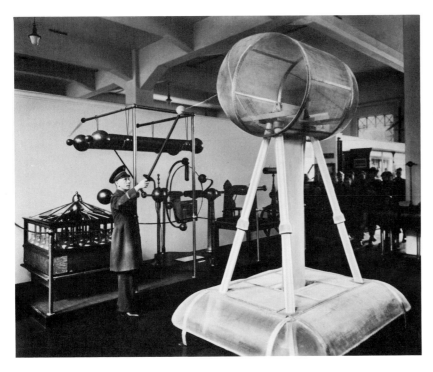

408. Deutsches Museum. Founded in 1903 by the engineer Oskar von Miller and opened in 1925, this Munich museum was a model for the Museum of Science and Industry in Chicago and other institutions. The Deutsches Museum was almost completely destroyed by bombing during World War II, but it was rebuilt with spectacular new exhibits. The photograph, taken at mid-century, shows a portion of the physics department. In the right foreground is a Van de Graaff generator built in 1951. At left are Leyden jars and an electrical machine that belonged to the German physicist Georg Simon Ohm, dating from around 1835.

409. Patent drawing for the first planetarium projector. In 1913 the Deutsches Museum's Oskar von Miller asked the Carl Zeiss Optical Works to build a device to project images of the nighttime sky on the interior of a dome. As the schematic suggests, the projector was a complicated instrument. The sphere, with cone-shaped lens supports, was for star projections; the structure on the left housed separate projectors for the sun, the moon, and five planets. The planetarium could show about 4,500 stars, but only for the latitude of Munich. Later planetariums featured adjustable latitude.

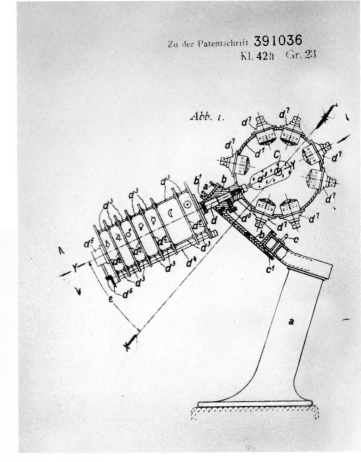

Zu der Patentschrift **391036**
Kl. 42h Gr. 23

Abb. 1.

410. Zeiss planetarium installed.
The Deutsches Museum unveiled the first Zeiss planetarium in October 1923. The stars were projected from glass plates similar to lantern slides. The device represented the geocentric motions of the planets to within a single degree, except for Mercury. The large black "saucer" is a gravity-controlled screen to provide a common horizon for the 31 conical star projectors.

411. Copernican ceiling orrery.
Also at the Deutsches Museum was
this older type of planetarium. A
set of overhead rails guided the
planets, from Mercury to Neptune,
in elliptical orbits. By riding in the
mobile cage under the model earth,
the heliocentric system could be
viewed from an earth-centered perspective. Large but simpler ceiling
orreries, without the mobile cage,
were later installed at the Hayden
Planetarium in New York and the
Morehead Planetarium in North
Carolina.

412. Lecture demonstration at the Adler Planetarium. The Zeiss instrument in the Deutsches Museum proved its worth as an educational device,
and the company proceeded to develop its more versatile "universal planetarium." By the outbreak of World War II, Zeiss had installed 25
machines in major cities such as Tokyo, Moscow, and Paris. This photograph at Chicago's Adler Planetarium was taken around 1935.

413. St. Louis World's Fair, 1904. Progress in science and technology has traditionally been a major theme of world's fairs or expositions. The Louisiana Purchase Exposition in St. Louis was no exception. Here we see the exhibit of the American De Forest Wireless Company. The fair's focus on science was not limited to exhibitions, as scientists gathered from around the world to report on their latest discoveries.

414. Chicago World's Fair, 1893. Astronomical exhibits were featured in the corner of one balcony in the enormous Manufactures and Liberal Arts Building at the World's Columbian Exposition. In the background, rising high from the floor below, is the mounting for the new 40-inch telescope, built by the Warner & Swasey Company. Scientific entrepreneur George Ellery Hale, who had ordered the refractor for the Yerkes Observatory, organized an international astronomical congress in conjunction with the exposition.

415. Brussels, Expo '58. Dominating the grounds at the Brussels Universal and International Exposition in 1958 was the Atomium, a structure 100 meters (more than 300 feet) high consisting of nine spheres representing an alpha iron crystal magnified 160 billion times. A theme of the fair—the first international exposition following World War II—was the peaceful use of atomic energy, and here was one of the first public displays of a nuclear reactor.

416. Chicago, Century of Progress, 1933. Held forty years after the World's Columbian Exposition, the second Chicago fair had its lights turned on by a telescope and detector pointed to Arcturus, whose light had left the giant first-magnitude star forty years earlier. Scientific themes were abundant, especially in this Hall of Science. On the fairgrounds was the new Adler Planetarium, opened just three years earlier.

417. Milestones of Flight. One of the most striking exhibition rooms in all the world's museums is the Milestones of Flight gallery at the National Air and Space Museum in Washington, D.C. On display are the Wright brothers' first successful plane (*lower right*), a Mercury space capsule (*floor near the entrance door*), the X-1 that first broke the sound barrier (*upper center*), and Charles Lindbergh's *Spirit of St. Louis* (*upper left*). Other exhibitions include the Gemini and Apollo space capsules.

418. Soviet space program, Expo '58. A focal point of the Soviet building at the 1958 Brussels fair, next to an enormous statue of Lenin, was a display of the Soviets' rapidly developing space program. Here tens of thousands of visitors saw models of the first Sputniks.

419. Paris, 1937. The Caudron-Renault plane with which Lieutenant Michel Detroyat had won the Thompson Trophy in the United States in 1936 took the place of honor in the Palais de l'Air at the International Exposition of Arts and Techniques in Modern Life. The pavilion, designed by artists Felix Aublet and Robert Delaunay, pioneered the use of glass walls and featured French aircraft and engines. Note the visitor passage spiraling around Detroyat's airplane in the center of the hall.

420. Soviet space station, Expo 86. The theme of the 1986 world's fair in Vancouver, Canada, was transportation and communications. The Soviet pavilion celebrated the Soviet Union's pioneering role in space exploration and featured a huge statue of Yuri Gagarin and a duplicate of the Salyut space station that was orbiting overhead. In the foreground is a Soyuz spacecraft in docking position.

Nobel Laureates in Physics and Chemistry

PHYSICS

1901 Wilhelm Konrad Röntgen (Roentgen) (1845–1923), German. Discovery of X rays.

1902 Hendrik Antoon Lorentz (1853–1928), Dutch, and **Pieter Zeeman** (1865–1943), Dutch. Research on the influence of magnetism on radiation phenomena.

1903 Antoine Henri Becquerel (1852–1908), French. Discovery of spontaneous radioactivity. **Pierre Curie** (1859–1906), French, and **Marie Curie** (b. Skłodowska) (1867–1934), French. Joint research on the radiation phenomena discovered by Becquerel.

1904 Lord Rayleigh (John William Strutt) (1842–1919), British. Investigation of gases and discovery of argon.

1905 Philipp Eduard Anton von Lenard (1862–1947), German. Research on cathode rays.

1906 Joseph John Thomson (1856–1940), British. Research on the conduction of electricity by gases.

1907 Albert Abraham Michelson (1852–1931), American. Precision spectroscopic and metrological investigations.

1908 Gabriel Lippmann (1845–1921), French. Photographic color reproduction based on the phenomenon of interference.

1909 Guglielmo Marconi (1874–1937), Italian, and **Karl Ferdinand Braun** (1850–1918), German. Contributions to the development of wireless telegraphy.

1910 Johannes Diderik van der Waals (1837–1923), Dutch. Work on the equation of state for gases and liquids.

1911 Wilhelm Wien (1864–1928), German. Discoveries regarding the laws of heat radiation.

1912 Nils Gustaf Dalén (1869–1937), Swedish. Invention of automatic regulators for illuminating lighthouses and buoys.

1913 Heike Kamerlingh Onnes (1853–1926), Dutch. Investigations of the properties of matter at low temperatures.

1914 Max von Laue (1879–1960), German. Discovery of the diffraction of X rays by crystals.

1915 Sir William Henry Bragg (1862–1942), British, and his son **Sir William Lawrence Bragg** (1890–1971), British. Analysis of crystal structure by means of X rays.

1916 No award.

1917 Charles Glover Barkla (1877–1944), British. Discovery of the characteristic Roentgen radiation of the elements.

1918 Max Planck (1858–1947), German. Discovery of energy quanta.

1919 Johannes Stark (1874–1957), German. Discovery of the Doppler effect in canal rays and the splitting of spectral lines in electric fields.

1920 Charles Édouard Guillaume (1861–1938), French. Discovery of anomalies in nickel steel alloys.

1921 Albert Einstein (1879–1955), German. Services to theoretical physics, especially discovery of the law of the photoelectric effect.

1922 Niels Henrik David Bohr (1885–1962), Danish. Investigation of the structure of atoms and their radiation.

1923 Robert Andrews Millikan (1868–1953), American. Work on the elementary charge of electricity and on the photoelectric effect.

1924 Karl Manne Georg Siegbahn (1886–1978), Swedish. Discoveries in X-ray spectroscopy.

1925 James Franck (1882–1964), German, and **Gustav Hertz** (1887–1975), German. Discovery of the laws governing the impact of an electron on an atom.

1926 Jean Baptiste Perrin (1870–1942), French. Discovery of sedimentation equilibrium; work on discontinuous structure of matter.

1927 Arthur Holly Compton (1892–1962), American. Discovery of the Compton effect. **Charles Thomson Rees Wilson** (1869–1959), British.

421. Schrödinger (1933)

Method of making the paths of electrically charged particles visible by condensation of vapor.

1928 Owen Willans Richardson (1879–1959), British. Work on the thermionic phenomenon and discovery of Richardson's law.

1929 Louis Victor de Broglie (1892–1987), French. Discovery of the wave nature of electrons.

1930 Sir Chandrasekhara Venkata Raman (1888–1970), Indian. Discovery of Raman effect and work on the scattering of light.

1931 No award.

1932 Werner Heisenberg (1901–1976), German. Creation of quantum mechanics.

1933 Erwin Schrödinger (1887–1961), Austrian, and **Paul Adrien Maurice Dirac** (1902–1984), British. Work in atomic theory.

1934 No award.

1935 James Chadwick (1891–1974), British. Discovery of the neutron.

1936 Victor Franz Hess (1883–1964), Austrian-American. Discovery of cosmic radiation. **Carl David Anderson** (1905–   ), American. Discovery of the positron.

1937 Clinton Joseph Davisson (1881–1958), American, and **Sir**

George Paget Thomson (1892–1975), British. Discovery of the diffraction of electrons by crystals.

1938 Enrico Fermi (1901–1954), Italian. Discovery of nuclear reactions brought about by slow neutrons, and demonstrations of the existence of new radioactive elements produced by neutron irradiation.

1939 Ernest Orlando Lawrence (1901–1958), American. Invention and development of the cyclotron.

1940 No award.

1941 No award.

1942 No award.

1943 Otto Stern (1888–1969), German-American. Discovery of the magnetic moment of the proton and contribution to the development of the molecular ray method.

1944 Isidor Isaac Rabi (1898–1988), American. Resonance method for recording magnetic properties of atomic nuclei.

1945 Wolfang Pauli (1900–1958), Austrian. Discovery of the exclusion principle, or Pauli principle.

1946 Percy Williams Bridgman (1882–1961), American. Inventions and discoveries in the field of high-pressure physics.

1947 Sir Edward Victor Appleton (1892–1965), British. Discovery of the Appleton layer and investigations of the upper atmosphere.

1948 Patrick Maynard Stuart Blackett (1897–1974), British. Development of the Wilson cloud-chamber method.

1949 Hideki Yukawa (1907–1981), Japanese. Prediction of the existence of mesons.

1950 Cecil Frank Powell (1903–1969), British. Development of the photographic method of studying nuclear processes.

1951 Sir John Douglas Cockcroft (1897–1967), British, and **Ernest Thomas Sinton Walton** (1903–   ),

422. Shockley (*seated*), Bardeen, and Brattain (1956)

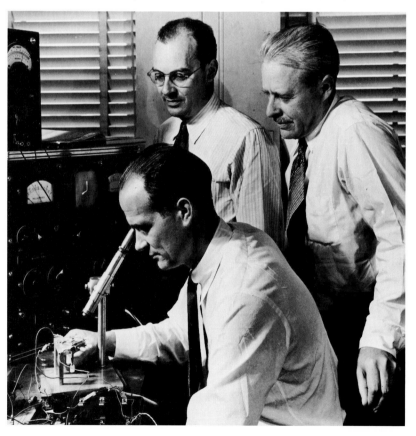

Irish. Work on transmutation of atomic nuclei by artificially accelerated atomic particles.

1952 Felix Bloch (1905–1983), American, and **Edward Mills Purcell** (1912–   ), American. Development of new methods for nuclear magnetic precision measurements and related discoveries.

1953 Frits (Frederik) Zernike (1888–1966), Dutch. Invention of the phase-contrast microscope.

1954 Max Born (1882–1970), German-British. Statistical interpretation of the wave function and research in quantum mechanics. **Walther Bothe** (1891–1957), German. Coincidence method.

1955 Willis Eugene Lamb (1913–   ), American. Discoveries in the hydrogen spectrum. **Polykarp Kusch** (1911–   ), German-American. Determination of the magnetic moment of the electron.

1956 William Shockley (1910–   ), American, **John Bardeen** (1908–   ), American, and **Walter Houser Brattain** (1902–1987), American. Discovery of the transistor effect.

1957 Chen Ning Yang (1922–   ), Chinese-American, and **Tsung Dao Lee** (1926–   ), Chinese-American. Work leading to the discovery of violations of left-right symmetry in physical processes.

1958 Pavel A. Cherenkov (1904–   ), Soviet, **Ilya M. Frank** (1908–   ), Soviet, and **Igor E. Tamm** (1895–1971), Soviet. Discovery and interpretation of the Cherenkov effect.

1959 Owen Chamberlain (1920–   ), American, and **Emilio Segrè** (1905–   ), Italian-American. Discovery of the antiproton.

1960 Donald A. Glaser (1926–   ), American. Invention of bubble chamber for studying elementary particles.

1961 Robert Hofstadter (1915–   ), American. Discoveries concerning the nucleus of the atom. **Rudolf L. Mössbauer** (1929–   ), German. Work on resonance absorption of gamma radiation.

1962 Lev Davydovich Landau (1908–1968), Soviet. Pioneering theories on condensed matter, especially liquid helium.

1963 Eugene Paul Wigner (1902–   ), Hungarian-American. Formulation of laws of symmetry governing interactions of nuclear particles. **J. Hans D. Jensen** (1906–1973), German, and **Maria Mayer** (b. Goeppert) (1906–1972), German-American. Research into the shell structure of the atomic nucleus.

1964 Charles Hard Townes (1915–   ), American, **Nikolai Gennadievich Basov** (1922–   ), Soviet, and **Aleksandr Mikhailovich Prokhorov** (1916–   ), Soviet. Development of maser-laser principle of amplifying electromagnetic radiation.

1965 Richard Phillips Feynman (1918–1988), American, **Julian Seymour Schwinger** (1918–   ), American, and **Shinichiro Tomonaga** (1906–1979), Japanese. Study of subatomic particles.

1966 Alfred Kastler (1902–1984), French. Study of atomic structure by use of radiation.

1967 Hans Albrecht Bethe (1906–   ), American. Discoveries concerning the energy production of stars.

1968 Luis W. Alvarez (1911–1988), American. Contributions to the physics and detection of elementary particles.

1969 Murray Gell-Mann (1929–   ), American. Discoveries concerning elementary particles.

1970 Hannes Alfvén (1908–   ), Swedish. Work in magnetohydrodynamics. **Louis Néel** (1904–   ), French. Discoveries concerning antiferromagnetism and ferrimagnetism.

1971 Dennis Gabor (1900–1979), British. Invention of holography, a lensless system of three-dimensional photography.

1972 John Bardeen (1908–   ), American, **Leon N. Cooper** (1930–   ), American, and **John R. Schrieffer** (1931–   ), American. Theory of superconductivity.

1973 Leo Esaki (1925–   ), Japanese-American, and **Ivar Giaever** (1929–   ), Norwegian-American. Work with semiconductors and superconductors. **Brian D. Josephson** (1940–   ), British. Theoretical predictions of the properties of supercurrent through a tunnel barrier.

423. P. W. Anderson (1977)

1974 Antony Hewish (1924–   ), British, and **Sir Martin Ryle** (1918–1984), British. Pioneering work in radio astrophysics.

1975 Aage Bohr (1922–   ), Danish, **Ben R. Mottelson** (1926–   ), American-Danish, and **James Rainwater** (1919–1986), American. Discovery of the connection between collective motion and particle motion in the atomic nucleus, and the development of the theory of the structure of the atomic nucleus based on this link.

1976 Burton Richter (1931–   ), American, and **Samuel C. C. Ting** (1936–   ), American. Discovery of the J, or psi, particle.

1977 Philip W. Anderson (1924–   ), American, **John H. Van Vleck** (1899–1980), American, and **Sir Nevill F. Mott** (1905–   ), British. Work underlying the development of computer memories, copying machines, and other electronic devices.

1978 Arno Penzias (1933–   ), German-American, and **Robert Wilson** (1936–   ), American. Discovery of cosmic microwave background radiation. **Pyotr Kapitsa** (1894–1984),

Soviet. Basic research into low-temperature physics.

1979 **Sheldon L. Glashow** (1933–), American, **Abdus Salam** (1926–), Pakistani, and **Steven Weinberg** (1933–), American. Theory of unity of electromagnetic and weak atomic forces.

1980 **James W. Cronin** (1931–), American, and **Val L. Fitch** (1923–), American. Solution of an apparent discrepancy between two laws of physics concerning the existence of matter.

1981 **Kai Siegbahn** (1918–), Swedish. Development of electron spectroscopy. **Nicolaas Bloembergen** (1920–), Dutch-American, and **Arthur Leonard Schawlow** (1921–), American. Development of laser spectroscopy.

1982 **Kenneth Geddes Wilson** (1936–), American. Analysis of basic changes in states of matter.

1983 **Subrahmanyan Chandrasekhar** (1910–), Indian-American, and **William A. Fowler** (1911–), American. Research into the origin, evolution, and composition of stars.

1984 **Carlo Rubbia** (1934–), Italian, and **Simon van der Meer** (1925–), Dutch. Discovery of three elementary particles, confirming theory of the unity of the weak and electromagnetic forces.

1985 **Klaus von Klitzing** (1943–), West German. Discovery of a technique for accurately describing electrical conductivity in terms of quantum theory.

1986 **Ernst Ruska** (1906–1988), West German. Invention of electron microscope. **Gerd Binnig** (1947–), West German, and **Heinrich Rohrer** (1933–), Swiss. Development of scanning tunneling microscope.

1987 **Karl Alex Müller** (1927–), Swiss, and **J. Georg Bednorz** (1950–), West German. Discovery of high-temperature superconductivity in ceramics.

1988 **Leon Lederman** (1922–), American, **Melvin Schwartz** (1932–), American, and **Jack Steinberger** (1921–), American. Method for producing beam of high-energy, high- intensity neutrinos and discovery of the muon neutrino.

CHEMISTRY

1901 **Jacobus Henricus van't Hoff** (1852–1911), Dutch. Discovery of the laws of chemical dynamics and osmotic pressure in solutions.

1902 **Emil Hermann Fischer** (1852–1919), German. Work on sugar and purine syntheses.

1903 **Svante August Arrhenius** (1859–1927), Swedish. Electrolytic theory of dissociation.

1904 **Sir William Ramsay** (1852–1916), British. Discovery of the inert gaseous elements in air.

1905 **Adolf von Baeyer** (1835–1917), German. Work on organic dyes and hydroaromatic compounds.

1906 **Henri Moissan** (1852–1907), French. Isolation of fluorine and development of the Moissan electric furnace.

1907 **Eduard Buchner** (1860–1917), German. Biochemical researches and discovery of cell-free fermentation.

1908 **Ernest Rutherford** (1st Baron Rutherford of Nelson) (1871–1937), British. Work on the disintegration of elements and the chemistry of radioactive substances.

1909 **Wilhelm Ostwald** (1853–1932), German. Work on catalysis and the principles of chemical equilibria and rates of reaction.

1910 **Otto Wallach** (1847–1931), German. Pioneer work in the field of alicyclic compounds.

1911 **Marie Curie** (b. Skłodowska) (1867–1934), French. Discovery of radium and polonium and isolation and study of radium.

1912 **Victor Grignard** (1871–1935), French. Discovery of the Grignard reagent. **Paul Sabatier** (1854–1941), French. Method of hydrogenating organic compounds.

1913 **Alfred Werner** (1866–1919), Swiss. Work on the linkage of atoms in molecules.

1914 **Theodore William Richards** (1868–1928), American. Determination of atomic weight of elements.

1915 **Richard Martin Willstätter** (1872–1942), German. Research on structure of plant pigments, especially chlorophyll.

1916 No award.

1917 No award.

1918 **Fritz Haber** (1868–1934), German. Synthesis of ammonia.

1919 No award.

1920 **Walther Hermann Nernst** (1864–1941), German. Work in thermochemistry.

1921 **Frederick Soddy** (1877–1956), British. Research in the chemistry of radioactive substances and investigation of isotopes.

1922 **Francis William Aston** (1877–1945), British. Discovery of isotopes in nonradioactive elements with mass spectrograph and enunciation of the whole-number rule.

1923 **Fritz Pregl** (1869–1930), Austrian. Invention of the method of microanalysis of organic substances.

1924 No award.

1925 **Richard Adolf Zsigmondy** (1865–1929), German. Demonstration of the heterogeneous nature of colloid solutions.

1926 **Theodor Svedberg** (1884–1971), Swedish. Work on disperse systems.

1927 **Heinrich Otto Wieland** (1877–1957), German. Work on the constitution of bile acids and related substances.

1928 **Adolf Otto Reinhold Windaus** (1876–1959), German. Research on sterols and their connection with vitamins.

1929 **Arthur Harden** (1865–1940), British, and **Hans Karl August Simon von Euler-Chelpin** (1873–1964), Swedish. Work on fermentation of sugar and fermentative enzymes.

1930 **Hans Fischer** (1881–1945), German. Research into constitution of hemin and chlorophyll and synthesis of hemin.

1931 **Carl Bosch** (1874–1940), German, and **Friedrich Bergius** (1884–1949), German. Work on chemical high-pressure methods.

1932 **Irving Langmuir** (1881–1957), American. Discoveries and investigations in surface chemistry.

1933 No award.

1934 **Harold Clayton Urey** (1893–1981), American. Discovery of heavy hydrogen.

1935 **Frédéric Joliot-Curie** (1900–1958), French, and **Irène Joliot-Curie** (1897–1956), French. Synthesis of new radioactive elements.

1936 Peter Joseph Wilhelm Debye (1884–1966), Dutch-American. Investigations on dipole moments and on the diffraction of X rays and electrons in gases.

1937 Walter Norman Haworth (1883–1950), British. Investigations of carbohydrates and vitamin C. **Paul Karrer** (1889–1971), Swiss. Work on carotenoids, flavins and vitamins A and B$_2$.

1938 Richard Kuhn (1900–1967), German. Research in carotenoids and vitamins.

1939 Adolf Friedrich Johann Butenandt (1903–   ), German. Sex hormones. **Leopold Ruzicka** (1887–1976), Swiss. Work on polymethylenes and higher terpenes.

1940 No award.

1941 No award.

1942 No award.

1943 George de Hevesy (1885–1966), Hungarian. Use of isotopes as tracers in the study of chemical processes.

1944 Otto Hahn (1879–1968), German. Discovery of the fission of heavy nuclei.

1945 Artturi Ilmari Virtanen (1895–1973), Finnish. Fodder-preservation and work in agricultural and nutrition chemistry.

1946 James Batcheller Sumner (1887–1955), American. Discovery that enzymes can be crystallized. **John Howard Northrop** (1891–1987), American, and **Wendell Meredith Stanley** (1904–1971), American. Preparation of enzymes and virus proteins in pure form.

1947 Sir Robert Robinson (1886–1975), British. Investigations on plant products, especially the alkaloids.

1948 Arne Wilhelm Kaurin Tiselius (1902–1971), Swedish. Research on electrophoresis, adsorption analysis, and serum proteins.

1949 William Francis Giauque (1895–1982), American. Work on chemical thermodynamics.

1950 Otto Paul Hermann Diels (1876–1954), German, and **Kurt Alder** (1902–1958), German. Discovery of the diene synthesis.

1951 Edwin Mattison McMillan (1907–   ), American, and **Glenn Theodore Seaborg** (1912–   ),

424. Seaborg (1951) holding first sample of plutonium 239

American. Discoveries in the chemistry of transuranium elements.

1952 Archer John Porter Martin (1910–   ), British, and **Richard Laurence Millington Synge** (1914–   ), British. Invention of partition chromatography.

1953 Hermann Staudinger (1881–1965), German. Discoveries in macromolecular chemistry.

1954 Linus Carl Pauling (1901–   ), American. Research into the nature of the chemical bond.

1955 Vincent du Vigneaud (1901–1978), American. Work on sulfur compounds and first synthesis of a polypeptide hormone.

1956 Sir Cyril Norman Hinshelwood (1897–1967), British, and **Nikolai Nikolaevich Semenov** (1896–1986), Soviet. Researches into the mechanism of chemical reactions.

1957 Sir Alexander R. Todd (1907–   ), British. Work on nucleotides and nucleotide coenzymes.

1958 Frederick Sanger (1918–   ), British. Work on the structure of proteins, especially insulin.

1959 Jaroslav Heyrovský (1891–1967), Czechoslovakian. Development of polarography, an electrochemical method of analysis.

1960 Willard F. Libby (1908–1980), American. Development of a method of using radioactive carbon to determine age of objects.

1961 Melvin Calvin (1911–   ), American. Discoveries concerning photosynthesis.

1962 John Kendrew (1917–   ), British, and **Max Perutz** (1914–   ), British-American. Studies of globular proteins.

1963 Karl Ziegler (1898–1973), German, and **Giulio Natta** (1903–1979), Italian. Development of a system to control polymerization of simple hydrocarbons into multimolecular compounds.

1964 Dorothy Crowfoot Hodgkin (1915–   ), British. Determination of the structure of biochemical compounds essential in combating pernicious anemia.

1965 Robert Burns Woodward (1917–1979), American. Synthesis of

425. Prigogine (1977)

complex organic compounds, including chlorophyll.

1966 Robert S. Mulliken (1896–1986), American. Work on the chemical bond and electronic structure of molecules.

1967 Manfred Eigen (1927–), West German, **Ronald G. W. Norrish** (1897–1978), British, and **George Porter** (1920–), British. Studies of extremely fast chemical reactions.

1968 Lars Onsager (1903–1976), American. Discovery of the reciprocal relations fundamental for the thermodynamics of irreversible processes.

1969 Derek H. R. Barton (1918–), British, and **Odd Hassel** (1897–1981), Norwegian. Work on the shape of organic molecules.

1970 Luis F. Leloir (1906–1987), French-Argentinian. Discovery of sugar nucleotides and their role in biosynthesis of carbohydrates.

1971 Gerhard Herzberg (1904–), Canadian. Research on the structure of the molecule.

1972 Christian B. Anfinsen (1916–), American, **Stanford Moore** (1913–1982), American, and **William**

H. Stein (1911–1980), American. Major contributions to enzyme chemistry.

1973 Ernst O. Fischer (1918–), German, and **Geoffrey Wilkinson** (1921–), British. Pioneering work on the chemistry of the organometallic "sandwich compounds."

1974 Paul J. Flory (1910–1985), American. Analytic methods to study the properties and molecular architecture of long chain molecules.

1975 John W. Cornforth (1917–), Australian-British, and **Vladimir Prelog** (1906–), Czech-Swiss. Contributions to stereochemistry.

1976 William N. Lipscomb, Jr. (1920–), American. Studies on the structure of boranes, illuminating problems of chemical bonding.

1977 Ilya Prigogine (1917–), Belgian. Contributions to nonequilibrium thermodynamics, notably the theory of dissipative structures.

1978 Peter Mitchell (1920–), British. Studies in bioenergetics.

1979 Herbert C. Brown (1912–), American, and **Georg Wittig** (1897–1987), West German. Development of temporary chemical links for complex molecules.

1980 Paul Berg (1926–), American, **Walter Gilbert** (1932–), American, and **Frederick Sanger** (1918–), British. Methods of determining in detail the structure and function of DNA.

1981 Kenichi Fukui (1918–), Japanese, and **Roald Hoffman** (1937–), Polish-American. Application of quantum mechanics to predict the course of chemical reactions.

1982 Aaron Klug (1926–), expatriate South African. Development of crystallographic electron microscopy, particularly in connection with the structure of viruses.

1983 Henry Taube (1915–), Canadian-American. Research in the transfers of electrons in metals through chemical reactions.

1984 R. Bruce Merrifield (1921–), American. Automated method of assembling peptides to synthesize proteins.

1985 Herbert A. Hauptman (1917–), American, and **Jerome Karle**

(1918–), American. Mathematical methods for determining from X-ray diffraction patterns the structure of hormones and other biological molecules.

1986 Dudley R. Herschbach (1932–), American, **Yuan T. Lee** (1936–), Chinese-American, and **John C. Polanyi** (1929–), Canadian. Work on reaction dynamics, using physical theory and technology to illustrate how energy shapes chemical reactions.

1987 Donald J. Cram (1919–), American, **Charles J. Pedersen** (1904–), Norwegian-American, and **Jean-Marie Lehn** (1939–), French. Work on synthetic molecules that can mimic vital chemical reactions of life processes.

1988 Johann Deisenhofer (1943–), West German, **Robert Huber** (1937–), West German, and **Hartmut Michel** (1948–), West German. Determination of structure of a bacterial protein that performs simple photosynthesis.

426. Klug (1982)

Guide to Further Reading

GENERAL BACKGROUND

Darius, Jon. *Beyond Vision*. Oxford: Oxford University Press, 1984.

Gillispie, Charles Coulston, ed. *Dictionary of Scientific Biography*. New York: Scribners, 1970—. 15 vols. plus supplements. This set is consistently one of the chief sources for information on twentieth-century science.

Leprince-Ringuet, Louis, ed. *Les Inventeurs célèbres: sciences physiques et applications*. Paris: L. Mazenod, 1950. In German: *Berühmten Erfinder: Physiker und Ingenieure*. Cologne: Aulis Verlag, 1965.

Schneer, Cecil J. *Search for Order: The Development of the Major Ideas in the Physical Sciences From the Earliest Times to the Present*. New York: Harper, 1960. Reprint: *The Evolution of Physical Science: Major Ideas From Earliest Times to the Present*. Lanham, Md.: University Press of America, 1984.

Wilson, Mitchell. *American Science and Invention: A Pictorial History*. New York: Simon and Schuster, 1954.

PART I: THE AGE OF RUTHERFORD, EINSTEIN, AND BOHR

Chapter 1: Radioactivity and the Nucleus

Andrade, E.N. da C. *Rutherford and the Nature of the Atom*. Garden City: Doubleday, 1964.

Curie, Eve. *Madame Curie*. Vincent Sheean, trans. Garden City: Doubleday, Doran and Company, 1938.

Weinberg, Steven. *Discovery of Subatomic Particles*.

Scientific American Library. New York: W. H. Freeman, 1983.

Wilson, David. *Rutherford, Simple Genius*. Cambridge, Mass.: MIT Press, 1983.

Chapter 2: Atomic Chemistry

Heilbron, John L. *H. G. J. Moseley: The Life and Letters of an English Physicist, 1887–1915*. Berkeley: University of California Press, 1974.

Mendelssohn, K. *The Quest for Absolute Zero: The Meaning of Low Temperature Physics*. New York: McGraw-Hill, 1966.

Stranges, Anthony N. *Electrons and Valence: Development of the Theory, 1900–1925*. College Station: Texas A&M University Press, 1982.

Thomson, George. *J. J. Thomson and the Cavendish Laboratory in His Day*. Garden City: Doubleday, 1965.

Chapter 3: Relativity

Hoffmann, Banesh, and Helen Dukas. *Albert Einstein, Creator and Rebel*. New York: Viking, 1972.

Pais, Abraham. *Subtle Is the Lord: The Science and Life of Albert Einstein*. Oxford: Oxford University Press, 1982.

Smith, James H. *Introduction to Special Relativity*. New York: W. A. Benjamin, 1965.

Chapter 4: Quantum Mechanics

Cline, Barbara L. *Men Who Made a New Physics: Physicists and the Quantum Theory*. Chicago: University of Chicago Press, 1987. Revised edition of *The*

Questioners: Physicists and the Quantum Theory. New York: Crowell, 1964.

French, A. P., and P. J. Kennedy, eds. *Niels Bohr: A Centenary Volume.* Cambridge, Mass.: Harvard University Press, 1985.

Hoffmann, Banesh. *Strange Story of the Quantum: An Account for the General Reader of the Growth of the Ideas Underlying Our Present Atomic Knowledge.* 2nd ed. New York: Dover, 1959.

Keller, Alex. *The Infancy of Atomic Physics: Hercules in His Cradle.* New York: Clarendon Press–Oxford University Press, 1983.

———. "Henry Norris Russell." *Scientific American,* May 1989.

Verschuur, Gerrit L. *Interstellar Matters.* New York: Springer-Verlag, 1989.

PART II: THE STARS AND BEYOND

Chapter 5: Building Large Telescopes

King, Henry C. *The History of The Telescope.* Cambridge, Mass.: Sky Publishing, 1955.

Van Helden, Albert. "Building Large Telescopes, 1900–1950." *Astrophysics and Twentieth-Century Astronomy to 1950.* Ed. Owen Gingerich. Vol. 4A of *The General History of Astronomy.* Cambridge: Cambridge University Press, 1984. Pp. 134–152.

Wright, Helen. *Explorer of the Universe: A Biography of George Ellery Hale.* New York: Dutton, 1966.

Wright, Helen, Joan N. Warnow, and Charles Weiner. *The Legacy of George Ellery Hale.* Cambridge, Mass.: MIT Press, 1972.

Chapter 6: The Scale of the Universe

Berendzen, Richard, Richard Hart, and Daniel Seeley. *Man Discovers the Galaxies.* New York: Science History Publications, 1976.

Shapley, Harlow. *Through Rugged Ways to the Stars.* New York, Scribners, 1969.

Smith, Robert. *The Expanding Universe: Astronomy's "Great Debate" 1900–1931.* Cambridge: Cambridge University Press, 1982.

Whitney, Charles A. *The Discovery of Our Galaxy.* New York: Knopf, 1971.

Chapter 7: Stars and Interstellar Matter

DeVorkin, David. "Stellar Evolution and the Origin of the Hertzsprung-Russell Diagram." *Astrophysics and Twentieth-Century Astronomy to 1950.* Ed. Owen Gingerich. Vol. 4A of *The General History of Astronomy.* Cambridge: Cambridge University Press, 1984. Pp. 90–108.

PART III: THE EARTH AND ITS ENVIRONMENT

Chapter 8: The Age of the Earth

Cloud, Preston. *Cosmos, Earth, and Man: A Short History of the Universe.* New Haven: Yale University Press, 1978.

Eicher, Don L. *Geologic Time.* 2nd ed. Englewood Cliffs, N.J.: Prentice Hall, 1976.

Hallam, Anthony. *Great Geological Controversies.* New York: Oxford University Press, 1983.

Hurley, Patrick M. *How Old Is the Earth?* Garden City: Doubleday, 1959.

Chapter 9: Oceanography

Idyll, C. P., ed. *Exploring the Ocean World: A History of Oceanography.* Rev. ed. New York: Crowell, 1972.

Menard, H. W., ed. *Ocean Science.* Readings from Scientific American. San Francisco: W. H. Freeman, 1977.

Schlee, Susan. *The Edge of an Unfamiliar World: A History of Oceanography.* New York: Dutton, 1973.

———. "Man and the Oceans." *Wilson Quarterly,* summer 1984. Series of articles.

Chapter 10: Continental Drift

Gribben, John. *This Shaking Earth.* New York: G. P. Putnam's Sons, 1978.

Marvin, Ursula. *Continental Drift, the Evolution of a Concept.* Washington: Smithsonian Institution Press, 1973.

Miller, Russell. *Continents in Collision.* Alexandria, Va.: Time-Life Books, 1983.

Sullivan, Walter. *Continents in Motion.* New York: McGraw-Hill, 1974.

Chapter 11: Meteorology

Allen, Oliver E. *Atmosphere.* Alexandria, Va. Time-Life Books, 1983.

The Office of Charles & Ray Eames. *A Computer Perspective.* Cambridge, Mass.: Harvard University Press, 1973.

Vaeth, Joseph Gordon. *Weather Eyes in the Sky: America's Meteorological Satellites*. New York: Ronald Press, 1965.

Whipple, A. B. *Storm*. Alexandria, Va.: Time-Life Books, 1982.

PART IV: HARNESSING THE ATOM

Chapter 12: Nuclear Physics

Brown, Laurie M. "Hideki Yukawa and the Meson Theory." *Physics Today*, vol. 39, December 1986, pp. 55–62.

Davis, Nuel Pharr. *Lawrence and Oppenheimer*. New York: Simon & Schuster, 1968.

Fermi, Laura. *Atoms in the Family: My Life With Enrico Fermi*. Chicago: University of Chicago Press, 1954.

Weart, Spencer, and Melba Phillips, eds. *History of Physics*. New York: American Institute of Physics, 1985.

Chapter 13: Nuclear Fission and Nuclear Fusion

Badash, Lawrence, Joseph O. Hirschfelder, and Herbert P. Broida, eds. *Reminiscences of Los Alamos: 1943–1945*. Dordrecht: Reidel, 1980.

Blow, Michael, and the editors of *American Heritage*. *The History of the Atomic Bomb*. New York: American Heritage, 1968.

Goodchild, Peter. *J. Robert Oppenheimer, Shatterer of Worlds*. Boston: Houghton-Mifflin, 1981.

Rhodes, Richard. *The Making of the Atomic Bomb*. New York: Simon & Schuster, 1986.

Chapter 14: Inside the Nucleus

Brown, Laurie M., and Lillian Hoddeson, eds. *The Birth of Particle Physics*. Cambridge: Cambridge University Press, 1983.

Close, Frank, Michael Marten, and Christine Sutton. *The Particle Explosion*. New York: Oxford University Press, 1987.

Gallison, Peter Louis. *How Experiments End*. Chicago: University of Chicago Press, 1987.

Ne'eman, Yuval, and Yoram Kirsh. *The Particle Hunters*. Cambridge: Cambridge University Press, 1986.

Taubes, Gary. *Nobel Dreams: Power, Deceit, and the Ultimate Experiment*. New York: Random House, 1986.

PART V: THE STRUCTURE OF MATTER

Chapter 15: How Atoms Unite

Ihde, Aaron J. *The Development of Modern Chemistry*. New York: Harper & Row, 1964.

Mark, Herman F. *Giant Molecules*. New York: Time-Life Books, 1966.

Morris, Peter J. T. *Polymer Pioneers*. Philadelphia: Center for History of Chemistry, 1986.

Tarbell, Dean Stanley, and Ann Tracy Tarbell. *Essays on the History of Organic Chemistry in the United States, 1875–1955*. Nashville: Folio Publishers, 1986.

Thackray, Arnold, Jeffrey L. Sturchio, P. Thomas Carroll, and Robert Bud. *Chemistry in America, 1876–1976: Historical Indicators*. Dordrecht: Reidel, 1985.

Chapter 16: Chemical Technology

Furter, William F., ed. *A Century of Chemical Engineering*. New York: Plenum, 1982.

Haynes, Williams. *The American Chemical Industry*. 6 vols. New York: Van Nostrand, 1945–1954.

Kaufman, Morris. *The First Century of Plastics: Celluloid and Its Sequel*. London: Plastics Institute, 1963.

Taylor, F. Sherwood. *A History of Industrial Chemistry*. New York: Abelard Schuman, 1957.

PART VI: ELECTRONICS AND COMPUTERS

Chapter 17: Electronics

Antébi, Elizabeth. *Electronic Epoch*. New York: Van Nostrand-Reinhold, 1982.

Augarten, Stan. *State of the Art: A Photographic History of the Integrated Curcuit*. New York: Ticknor & Fields, 1983.

Braun, Ernst, and Stuart Macdonald. *Revolution in Miniature: The History and Impact of Semiconductor Electronics*. 2nd ed. Cambridge: Cambridge University Press, 1982.

Overhage, Carl F. J., ed. *The Age of Electronics*. New York: McGraw-Hill, 1962.

Chapter 18: The Computer Revolution

Augarten, Stan. *Bit by Bit: An Illustrated History of Computers and Their Inventors*. New York: Ticknor & Fields, 1984.

Goldstine, Herman. *The Computer from Pascal to Von Neumann*. Princeton: Princeton University Press, 1972.

The Office of Charles & Ray Eames. *A Computer Perspective*. Cambridge, Mass.: Harvard University Press, 1973.

Randall, Brian, ed. *The Origins of Digital Computers*. 3rd ed. New York: Springer-Verlag, 1982.

PART VII:
COSMIC VISTAS

Chapter 19: Radar and Radio Astronomy

Bowen, Edward G. *Radar Days*. Bristol: Adam Hilgar, 1987.

Hey, J. S. *The Evolution of Radio Astronomy*. New York: Science History Publications, 1973.

Sullivan, Woodruff T., III. "Early Radio Astronomy." *Astrophysics and Twentieth-Century Astronomy to 1950.* Ed. Owen Gingerich. Vol. 4A of *The General History of Astronomy*. Cambridge: Cambridge University Press, 1984. Pp. 190–198.

Swords, Sean S. *Technical History of the Beginnings of Radar*. London: Institute of Electrical Engineers, 1986.

Chapter 20: Men on the Moon

Brooks, Courtney G., James M. Grimwood, and Loyd S. Swenson, Jr. *Chariots for Apollo: A History of Manned Lunar Spacecraft*. Washington, D.C.: NASA, 1979.

Hall, R. Cargill. *Lunar Impact: A History of Project Ranger*. Washington, D.C.: NASA, 1977.

Von Braun, Wernher, and Frederick I. Ordway III. *History of Rocketry and Space Travel*. 3rd rev. ed. New York: Crowell, 1975.

Chapter 21: The Exploration of Space

Briggs, G. A., and F. W. Taylor. *The Cambridge Photographic Atlas of the Planets*. Cambridge: Cambridge University Press, 1982.

Compton, W. David, and Charles D. Benson. *Living and Working in Space: A History of Skylab*. Washington, D.C.: NASA, 1983.

Ezell, Edward Clinton, and Linda Neuman Ezell. *On Mars*. Washington, D.C.: NASA, 1984.

Newell, Homer E. *Beyond the Atmosphere: Early Years of Space Science*. Washington, D.C.: NASA, 1980.

Chapter 22: The Depths of the Universe

Marx, Siegfried, and Werner Pfau. *Observatories of the World*. New York: Van Nostrand-Reinhold, 1982.

Murdin, Paul, and David Allen. *Catalogue of the Universe*. Cambridge: Cambridge University Press, 1982.

Preston, Richard. "Beacons in Time: Maarten Schmidt and the Discovery of Quasars." *Mercury*, vol. 24, January-February 1988, pp. 2–11.

PART VIII:
THE SCIENCE OF SCIENCE

Chapter 23: Scientists Unite!

Kevles, Daniel J. *The Physicists: The History of a Scientific Community in Modern America*. New York: Borzoi–Knopf, 1977.

Sullivan, Walter. *Assault on the Unknown: The International Geophysical Year*. New York: McGraw-Hill, 1961.

Chapter 24: Communicating Science

King, Henry C., and J. R. Milburn. *Geared to the Stars: The Evolution of Planetariums, Orreries, and Astronomical Clocks*. Toronto: University of Toronto Press, 1978.

Rydell, Robert W. *All The World's a Fair: Visions of Empire at American International Expositions 1876–1916*. Chicago: University of Chicago Press, 1984.

———. "The Fan Dance of Science: American World's Fairs in the Great Depression." *Isis*, vol. 76, December 1985, pp. 525–542.

Tenner, Edward. "Pantheons of Nuts and Bolts." *American Heritage of Invention and Technology*, vol. 4, 1989, no. 3, pp. 16–22.

Picture Sources and Credits

All possible care has been taken to trace the ownership of each illustration and to make full acknowledgment for its use. If any errors or omissions have accidentally occurred, they will be corrected in subsequent editions provided notification is sent to the publisher. The following are the principal abbreviations used: AIP, American Institute of Physics; NASA, National Aeronautics and Space Administration.

1. NASA 69-H-1271. **2.** Jean Loup Charmet. **3.** AIP Niels Bohr Library. **4.** AIP Niels Bohr Library. **5.** AIP Niels Bohr Library. **6.** Smithsonian Institution, Neg. No. 76-16688. **7.** AIP Niels Bohr Library. **8.** Courtesy of J. Mishara, Museum Support Center, Smithsonian Institution. **9.** Emilio G. Segrè, "Nuclear Physics in Rome," in Robert H. Stuewer, ed., *Nuclear Physics in Retrospect: Proceedings of a Symposium on the 1930s*, Minneapolis, University of Minnesota Press, 1979, p. 48. **10.** Frederick Soddy, *Radio-Activity*, London, "The Electrician" Printing and Publishing Company, 1904, title page and p. 159; courtesy of Burndy Library. **11.** J. Rutherford, J. Chadwick, and C. D. Ellis, *Radiations From Radioactive Substances*, Cambridge, Cambridge University Press, 1930, plate X, fig. 1. **12.** Burndy Library. **13.** AIP Niels Bohr Library. **14.** J. B. Birks, ed., *Rutherford at Manchester*, New York, W. A. Benjamin, 1963, p. 70. **15.** Photograph by C. E. Wynn-Williams; courtesy of AIP Niels Bohr Library. **16.** *Philosophical Magazine*, vol. 21, 1911, pp. 669–688; courtesy of Burndy Library. **17.** AIP Niels Bohr Library. **18.** Courtesy of Harvard University Archives. **19.** (main picture) Emil Fischer papers, Bancroft Library, University of California at Berkeley. **19.** (right) A. P. French and P. J. Kennedy, eds., *Niels Bohr: A Centenary Volume*, Cambridge, Mass., Harvard University Press, 1985, p. 57. **20.** University of Cambridge, Department of Physics, Cavendish Laboratory. **21.** California Institute of Technology Archives. **22.** M. F. Perrin Archive; Louis Leprince-Ringuet, ed., *Les Inventeurs célèbres*,

Paris, L. Mazenod, 1950. **23.** *Philosophical Magazine*, vol. 26, 1913, plate XXIII. **24.** AIP Niels Bohr Library. **25.** AIP Niels Bohr Library. **26.** AIP Niels Bohr Library. **27.** Argonne National Laboratory. **28.** *Physical Review*, vol. 40, 1932, p. 9. **29.** Monde Laboratory of Physics, University of Cambridge. **30.** Kamerlingh Onnes Laboratory. **31.** *Communications From the Physical Laboratory at the University of Leiden*, no. 108, 1908, plate III; reproduced in George L. Trigg, *Landmark Experiments in Twentieth Century Physics*, New York, Crane-Russak, 1975, p. 41. **32.** Sovfoto. **33.** Lotte Jacobi Collection, Dimond Library, University of New Hampshire at Durham. **34.** (German) *Annalen der Physik*, vol. 17, 1905, p. 891. **34.** (English translation) H. A. Lorentz et al., *The Principle of Relativity*, trans. by W. Perett and G. B. Jeffrey, New York, Dodd, Mead, and Co., 1923, p. 37. **35.** Hale Observatories; courtesy of Burndy Library. **36.** D. C. Miller, *Reviews of Modern Physics*, vol. 5, 1933; courtesy of Burndy Library. **37.** AIP Niels Bohr Library. **38.** Royal Greenwich Observatory. **39.** Jewish National and University Library. **40.** National Museum of the History of Science, Leiden, Netherlands. **41.** René Magritte, *Time Transfixed*, 1938, oil on canvas, 147.0 x 98.7 cm, the Joseph Winterbotham Collection, 1970.426. © 1987 The Art Institute of Chicago. All Rights Reserved. **42.** DALI, Salvador. *The Persistence of Memory*. 1931. Oil on canvas, 9½ x 13". Collection, The Museum of Modern Art, New York. Given anonymously. **43.** Copyright Friedrich Meckseper. **44.** AIP Niels Bohr Library. **45.** *Fortune*, July 1932, p. 27. **46.** NASA 76-H-334. **47.** Hebrew University, Jerusalem. **48.** Hirshhorn Museum and Sculpture Garden, Smithsonian Institution. **49.** *Annalen der Physik*, vol. 4, 1901, p. 553. **50.** H. Rubens and F. Kurlbaum, *Annalen der Physik*, vol. 4, 1901, p. 649. **51.** Beckman Center for the History of Chemistry, University of Pennsylvania. **52.** Burndy Library. **53.** Government of Denmark/Stampazine. **54.** W. Grotrian, *Graphische Darstellung der Spektren*, Berlin, 1928. **55.** Niels Bohr Institute. **56.** AIP Niels Bohr Library. **57.** Spectroscopy Laboratory, Massachusetts Institute of Technology; courtesy of Owen Gingerich. **58.** AIP Niels Bohr Library. **59.** AIP Niels Bohr Library. **60.** Printed with the permission of the American Chemical Society. **61.** AIP Niels Bohr Library. **62.** Science Museum, London. **63.** AIP Niels Bohr Library. **64.** H. White, *Physical Review*, vol. 37, 1937, p. 1416. **65.** AIP Niels Bohr Library/ Uhlenbeck Collection. **66.** AIP Niels Bohr Library. **67.** U.S. Department of Commerce, National Bureau of Standards, Boulder, Colorado. **68.** AIP Niels Bohr Library; photograph courtesy of E. M. Purcell and N. Ramsay. **69.** William Vandivert; reproduced in "Photography by Laser," *Scientific American*, June 1965, p. 25. **70.** William Vandivert; reproduced in "Photography by Laser," *Scientific American*, June 1965, p. 26. **71.** AIP Niels Bohr Library. **72.** AIP Niels Bohr Library. **73.** AIP Niels Bohr Library. **74.** AIP Niels Bohr Library. **75.** AIP Niels Bohr Library. **76.** Mount Wilson and Las Campanas Observatories, Carnegie Institution of Washington; courtesy of Owen Gingerich. **77.** Mount Wilson and Las Campanas Observatories, Carnegie Institution of Washington. **78.** E. R. Hoge, Fairchild Aerial Surveys, Los Angeles; courtesy of AIP Niels Bohr Library. **79.** Hale Observatories; courtesy of AIP Niels Bohr Library. **80.** (top) Corning Glass Works; courtesy of AIP Niels Bohr Library. **80.** (bottom) Copyright California Institute of Technology, from Hansen Planetarium. **81.** Hale Observatories. **82.** AIP Niels Bohr Library. **83.** California Institute of Technology; reproduced in Henry C. King, *The History of the Telescope*, Cambridge, Mass., Sky Publishing Corporation, 1955, p. 374. **84.** Yerkes Observatory. **85.** Harvard College Observatory. **86.** Harvard College Observatory. **87.** *Harvard College Observatory Circular*, no. 173, 1912, p. 3. **88.** Lick Observatory. **89.** AIP Niels Bohr Library/ Shapley Collection. **90.** Yerkes Observatory. **91.** Yerkes Observatory. **92.** Mount Wilson/California Institute of Technolory. **93.** Adrian van Maanen, "Preliminary Evidence of Internal Motion in the Spiral Nebula Messier 101," *Astrophysical Journal*, vol. 44, 1916, pp. 210–228. **94.** Arthur Stanley Eddington, *Stellar Movements and the Structure of the Universe*, London, Macmillan, 1914, p. 97. **95.** J. S. Plaskett, *Dimensions and Structure of the Galaxy*, London, Oxford University Press, 1935, plate II; courtesy of AIP Niels Bohr Library. **96.** Harvard University; courtesy of Owen Gingerich. **97.** Hale Observatories; courtesy of AIP Niels Bohr Library. **98.** Hale Observatories; courtesy of AIP Niels Bohr Library. **99.** Hale Observatories; courtesy of AIP Niels Bohr Library; from Edwin Hubble, *The Realm of the Nebulae*, New Haven, Yale University Press, 1936, frontispiece. **100.** Hubble, *The Realm of the Nebulae*, plate vi; courtesy of AIP Niels Bohr Library. **101.** Lowell Observatory; courtesy of AIP Niels Bohr Library. **102.** Hubble, *The Realm of the Nebulae*, p. 114. **103.** Hubble, *The Realm of the Nebulae*, plate xiv. **104.** Hubble, *The Realm of the Nebulae*, plate viii. **105.** Yerkes Observatory. **106.** Mount Wilson and Las Campanas Observatories, Carnegie Institution of Washington. **107.** Henry Norris Russell, "Relations Between the Spectra and Other Characteristics of Stars," *Nature*, vol. 93, 1914, p. 252. **108.** Courtesy of E. Lamla, Astronomische Institut, Universität Bonn. **109.** Harvard College Observatory. **110.** *Annals of Harvard College Observatory*, vol. 28, 1901, pt. 2. **111.** Arthur Stanley Eddington, *The Internal Constitution of the Stars*, Cambridge, Cambridge University Press, 1926, p. 153. **112.** Lick Observatory. **113.** Yerkes Observatory. **114.** Whitney—Mount Wilson Observatory; courtesy of AIP Niels Bohr Library. **115.** Courtesy of Sacramento Peak Observatory, AURA. **116.** Hale Observatories; courtesy of AIP Niels Bohr Library. **117.** Ellerman—Palomar Observatory; courtesy of AIP Niels Bohr Library. **118.** *Sky & Telescope*. **119.** TASS from Sovfoto. **120.** (top) Griffith Observatory. **120.** (right) Mount Wilson and Palomar Observatories. **121.** Lowell Observatory photograph. **122.** The American Wilderness series: *The Grand Canyon*. Photograph by Harald Sund © 1972 Time-Life Books, Inc. **123.** Ursula Marvin, Smithsonian Astrophysical Laboratory. **124.** Smithsonian Astrophysical Laboratory. **125.** NASA 71-H-361. **126.** (top) NASA S-71-44836, Johnson Space Center Curator's Office, Houston, Texas. **126.** (right) NASA S-71-46588, Johnson Space Center Curator's Office, Houston, Texas. **127.** Dr. S. Moorbath; from Jim Brooks, *Origins of Life*, Tring, England, Lion Publishing Corporation, 1985, p. 114. **128.** Dr. Marjorie Muir/Jim Brooks; from Brooks, *Origins of Life*, p. 112. **129.** Preston Cloud, *Cosmos, Earth, and Man*, New Haven, Yale University Press, 1978, fig. 16. **130.** David Attenborough Productions Ltd. **131.** J. D. Macdougall, Scripps Institution of Oceanography; J. D. Macdougall, "Fission-Track Dating," *Scientific American*, December 1976, p. 115. **132.** USNC-IGY National Academy of Sciences photo. **133.** Chester C. Langway, Jr., SUNY at Buffalo. **134.** Courtesy, Field Museum of Natural History, Chicago. **135.** Official U.S. Navy photograph, USN 670256. **136.** New York Zoological Society photo. **137.** Vicky Culle, Woods Hole Oceanographic Institution. **138.** Woods Hole Oceanographic Institution. **139.** Woods Hole Oceanographic Institution and National Academy of Sciences, IGY Archives. **140.** Lawrence Sullivan, Lamont-Doherty Geological Obsevatory of Columbia University. **141.** IFREMER, Paris. **142.** John Donnelly, Woods Hole Oceanographic Institution. **143.** IFREMER, Paris. **144.** Scripps Institution of Oceanography. **145.** Teledyne Exploration Company. **146.** Kenneth J. Hsü, Swiss Federal Institute of Technology; from "When the Mediterranean Dried Up," *Scientific American*, December 1972, p. 28. **147.** NOAA, National Environment Satellite Data and Information Service. **148.** Global Marine Exploration Company. **149.** (left) Kenneth J. Hsü, Swiss Federal Institute of Technology; from "When the Mediterranean Dried Up," *Scientific American*, December 1972, p. 26. **149.** (right) Kenneth J. Hsü, Swiss Federal Institute of Technology; from *Scientific American*, December 1972, p. 35. **150.** TASS from Sovfoto. **151.** TASS from Sovfoto. **152.** National Geophysical Data Center. **153.** Alfred Wegener, *Die Entstehung der Kontinente und Ozeane*, Braunschweig, F. Vieweg & Sohn, 1922, pp. 4–5; used with permission of Friedrich Vieweg & Sohn. **154.** Allen Beechel; from Patrick M. Hurley, "The Confirmation of Continental Drift," *Scientific Amer-*

Index

The numbered references in this index refer either to textual material, which is designated by page numbers in lightface, or caption material, which is designated by illustration numbers in **boldface**. It is suggested that, in addition to the page and illustration number references given here for any particular subject, the reader also consult the Picture Sources and Credits section for any illustration listed, which may provide some additional information.

International Atomic Energy Agency, **400**
International Council of Scientific Unions, 260
International Exposition of Arts and Techniques in Modern Life, **419**
International Geophysical Year. *See* IGY
International Research Council, 260
international unions, scientific, 260
interstellar dust, 72, **112**
Io, **359**
Ioffe, Abram Feodorovich, **393**
ionic bond, 158
Iowa State University, 195
isotope, **10**, 14, 24, **25–28**

Jannasch, Holger, 92
Jansky, Karl, 209, **314**
Jeffreys, Harold, 102
Jensen, Johannes Hans Daniel, 283
Jodrell Bank radio telescope, 210, **321**
Joint European Torus (tokamak), **209**
Joliot-Curie, Frédéric and Irène, 124, 284
Jordan, Pascual, 34
Josephson, Brian David, 283
J/psi particle, 144, **221**
Jupiter, **362**

Kaiser Wilhelm Institute
  Chemistry, **192**
  Fiber Research, **51**
Kamerlingh Onnes, Heike, 14, **30**, **31**, 281, **394**
Kapitsa, Pyotr (Peter), **32**, 283
Kapteyn, Jacobus Cornelius, 71, **94**, 272
Karle, Jerome, 286
Karrer, Paul, 285
Kastler, Alfred, 283
Kazan physics conference, 1928, **393**
Keeler, James, 50, 71, **89**, 260
Kekulé, August, 157
Kelvin, Lord, 83
Kendrew, John, 285
Kernei, Otto, **263**
Kertész, André, **255**
Kilburn, Tom, **301**
Kilby, Jack, **284**

Kipfer, Paul, **167**
Kirchhoff, Gustav, 71
kite sounding, **173**
Kitt Peak National Observatory, 243, **371**, **372**
Klein, Oskar, **66**
Klitzing, Klaus von, 284
Klug, Aaron, 286, **426**
Knipping, Paul, **52**
Knoll, Max, **279**
Knott, Cargill Gilston, 101
Knudsen, Martin, **61**
Kola Peninsula (U.S.S.R.), deep hole, **150**, **151**
Komarov, Vladimir Mikhailovich, **331**
Kramers, Hendrik Anthony, 61, **66**
Kuhn, Richard, 285
Kuhn, Werner, 158, **236**
Kulik, Leonid Alekseevich, **118**, **119**
Kurchatov Institute of Atomic Energy (Moscow), **207**
Kusch, Polykarp, 283

laboratories
  astrophysics, **116**
  corporate, 182
  hot lab, **240**, **241**
  spectroscopy, 50
  See also specific individuals and institutions
Laika (dog in space), **329**
Lamb, Willis Eugene, Jr., **215**, 283
lambda-zero particle, **222**
Land, Edwin, **264**
Landau, Lev Davydovich, **66**, 283
Langevin, Paul, **61**
Langmuir, Irving, **61**, **272**, 284
laser, **69**
laser cooling, **244**
Laue, Max von, **51**, **52**, 281
Lawrence, Ernest Orlando, 124, **189–191**, **241**, 282
lawrencium, **241**
lead, in dating, 84
Leavitt, Henrietta, 59, **87**
Le Bel, Joseph Achille, 157
Lederman, Leon, 284
Lee, Tsung Dao, 283
Lee, Yuan Tseh, 286
Lehn, Jean-Marie, 286
Leiden Observatory, **319**
Leloir, Luis Frederico, 286

Lenard, Philipp Eduard Anton von, 281
Lewis, Gilbert Newton, 158, **231**
Libby, Willard Frank, **132**, 285
Lick, James, 49
Lick Observatory, 49, 50, **89**
  120-inch reflector, 243, **366**
life, history of, **123**, **128**, **130**
light, speed of, 23, 24, **34**, **68**
Lindblad, Bertil, 60
linear accelerator, 143, 144
Lippmann, Gabriel, 281
Lipscomb, William Nunn, Jr., 286
Little Boy, **201**
Lodge, Oliver, 181, 209, **270**
Loki (balloon-launched rocket), **343**
Lorentz, Hendrik Antoon, 23, **39**, **40**, **61**, 281, **390**, **394**
Lorenz, Edward, 112
Louisiana Purchase Exposition. *See* St. Louis World's Fair
Lovell, Bernard, 209, 210, **321**
Lowell, Percival, **101**, **354**
low-temperature physics, 14, **29**, **32**
Lulu, **138**
Luna (Soviet moon probe), **333**
Lunar Orbiter (U.S. moon probe), **335–337**
Lundin, Carl A. R., **74**
Lyell, Charles, 83

M13 (globular cluster), 59
M22 (globular cluster), 92
M31 (Andromeda galaxy), 59, 60, **97–100**, **318**, **386**
M81 (spiral nebula), **90**
M101 (spiral nebula), **93**
Maanen, Adriaan van, 93
Mack, Herman, **235**
magnetic-core memory, **302–304**
magnetic reversals, 84, 102, **156**
Magritte, René, **41**
manganese nodules, **139**
Manhattan Project, 133
Marchant, 195
Marconi, Guglielmo, 181, **265**, **269**, 281
Mariner (U.S. space probes), 350, **353**, **355**
Mark I (Harvard computer), 195, **296**
Mark I (Jodrell Bank radio telescope), **321**

orrery, Copernican ceiling, **411**
oscilloscope, 181, **265**
Ostwald, Wilhelm, 284
oxygen, liquid, **29**

Pacific, **144**, **155**
Pais, Abraham, **215**
Pallophotophone, **275**
Palomar Mountain (observatory)
    200-inch reflector, 49, 50, **79–81**,
      243
    Schmidt telescope, **83**
Paneth, Friedrich, **175**
Parícutin, **163**
particle accelerators, 143–144
    See also specific types
particle detectors, **7–9**, **11**, **12**, **13**,
    15, 143–144, **183**, **212**, **218**, **220**,
    224, **399**
particle physics, 123–124, 143–144,
    181–191, **211–229**
Pasteur, Louis, 157
Patterson, Clair, 84
Pauli, Wolfgang, 34, **61**, **66**, 123,
    143, 282
Pauling, Linus Carl, 158, **232**, 285
    nuclear test ban, 260, **402**
Pawsey, J. L., 210
Payne, Cecilia, 72
Pease, Francis, **71**
Pedersen, Charles J., 286
Peierls, Rudolf, 133
Pennsylvania, University of, 196, **289**
Penzias, Arno, 210, 283, **387**
periodicals, 259–260, **405**
periodic table, 4, 14, **19**, **23**
period-luminosity relation, **87**
Perrin, Jean Baptiste, 14, **22**, 281
Perutz, Max, 285
petroleum industry, 167, **252**, **254**
Philosophical Magazine, 259
Phobos, **358**
photoelectric effect, 33
photography, xi
    Polaroid, **264**
    Velox, 168, **257**
photomultiplier tube, **276**
photon, 124
Physical Review, 259, **405**
Physical Review Letters, **405**
physics, 281–284
    See also specific topics
Piccard, Auguste, **61**, 91, **135**, 167,
    168

Piccard, Jacques, **135**
Pickering, Edward Charles, 71
pillow lava, **140**
pion, 124, 143
Pioneer (U.S. space probes), 230,
    342, **363**, **364**
pitchblende, 3
Planck, Max, 33, **49**, **50**, **61**, 281, **394**
Planck's constant, 33, **49**, **68**
planetarium, 271–272, **409–412**
Plaskett, John Stanley, **95**
plastics, 168, **256–258**
plate tectonics, 102, **152**, **155**, 260
Pleiades, **112**
Plexiglas, 168, **259**
pliotron, **272**
Pluto, 72, **121**
plutonium, 133, **424**
Poincaré, Jules Henri, **394**
Polanyi, John Charles, 286
Polaroid, **264**
polonium, 3
polymers, 158, 167–168, **234–236**, 246
polystyrene, 168
Popov, Aleksandr Stepanovich, 181
Porter, George, 286
Porter, Russell W., **79**, **83**
positive rays, 25
positron, 123, **183**
Powell, Cecil Frank, 282
Pregl, Fritz, 284
Prigogine, Ilya, 286, **425**
Princeton Plasma Physics Labora-
    tory, **210**
Prokhorov, Aleksandr Mikhailovich,
    283
proton, **11**, 13
    lifetime, 144, **229**
Prout, William, 14
pulsar, 210, 244, **369**
Purcell, Edward Mills, 283, **317**
purines, **233**

quantum mechanics, 33–34, **48–50**,
    53–64, **67**, **68**
quark, 144, **221**, **222**
quasar, 210, 244, **377**, **378**

Rabi, Isidor Isaac, 158, 282, **311**
radar, 177, 209–210, **310–313**, **316**,
    351
Radiation Laboratory (MIT), **311**

radio, 181–182, **265**, **269**, **270**, **273**,
    274, 413
radioactivity, 4, 10
    artificial, **9**
    discovery, **2**, **3**, **5**
    fallout, 134
    protection against, **240**, **241**
radio astronomy, 209–210, 244, **317–**
    319
    telescopes, 209, **309**, **314–316**,
      320–324, **369**
radio galaxies, 210, **383–385**
radiometric dating, 84
    fission-track, **131**
    radiocarbon, **132**, **134**
radium, 3, **10**
Rainwater, James, 283
Raman, Chandrasekhara Venkata,
    282
Ramsay, William, **10**, 284
Ranger (U.S. moon probe), **334**
Ratcliffe, John Ashworth, **15**
Rayleigh, Lord, **281**
Reber, Grote, 209, **315**
Rebeur-Paschwitz, Ernst von, 102
rectifier, 181
redshift, **104**, **376–378**
Reines, Frederick, 143, **212**
relativity, 23–24, **33–36**, 60
    art, **41–43**
    black holes, 367
    quantum theory, **68**
    testing, **38–40**, 46
responder, 270
Richards, Theodore William, 14, **18**,
    284
Richardson, Lewis Fry, 112
Richardson, Owen Willans, **61**, 282
Richter, Burton, 283
Riemann, Georg Friedrich
    Bernhard, 24
Ritchey, George Willis, **90**
Robinson, Robert, 285
Rockefeller Foundation, 49, 50, **79**
rocketry, 219, 229, **325**, **327**, **343**,
    395
rockoon, 229
Röhm and Haas, 168, **259**
Rohrer, Heinrich, 284
Rome, 9
    1931 nuclear physics conference,
      388
Röntgen (Roentgen), Wilhelm Kon-
    rad, **3**, **4**, 281
Roosevelt, Franklin Delano, 133
Rose, Wickliffe, 49

Rossby, Carl-Gustaf, 92, 112, **171**
Royal Astronomical Society, 260
Royal Institution, 271
rubber, synthetic, 168, **253**, **255**
Rubbia, Carlo, 284
Runge, Carl, **389**
Ruska, Ernst, **279**, 284
Russell, Henry Norris, 71, 72, **107**, **108**
Rutherford, Ernest, 3–4, **7**, **10**, **11**, **13–17**, 83, 143, 284, **394**
Ruzicka, Leopold, 285
Ryle, Martin, 209, 210, 283, **324**

Sabatier, Paul, 284
Sagittarius, 60, **85**, 314
St. Louis World's Fair, 71, 272, **413**
Sakharov, Andrei Dmitrievich, 207
Salam, Abdus, 284
Salyut (Soviet space station), 332, **420**
San Andreas fault, **157**, **159**
San Francisco earthquake (1906), **159**, **160**
Sanger, Frederick, 285, 286
satellites, artificial earth
    meteorology, 112, **166**, **178**, **179**
    oceanography, 92, **147**
    tracking, **397**
    See also specific satellites
Saturn's rings, 230, **363**, **364**
Schawlow, Arthur Leonard, 284
Schmidt, Bernhard, **82**
Schmidt, Martin, **377**
Schmidt telescope, **82**, **83**
Schmitt, Harrison, 219, **340**
Schreyer, Helmut, **294**
Schrieffer, John Robert, 283
Schrödinger, Erwin, 34, **61**, **64**, 282, **421**
Schwartz, Melvin, 284
Schwinger, Julian Seymour, **215**, 283
Science Museum (London), **407**
scientific literature, proliferation of, **403–405**
Seaborg, Glenn Theodore, 285, **424**
seafloor spreading, 102, **156**
Segrè, Emilio, **9**, 144, 283
seismic profiling, 92, **145**, **146**
seismic sounding, Antarctic, **398**
seismograph, 101, 102, **161**
Seismological Society of Japan, 101
seismology, **158**

Semenov, Nikolai Nikolaevich, 285
semiconductors, 182
    See also chip; transistor
Serpukhov synchrotron, **227**
Shapley, Harlow, 50, 59, 60, **85**, **87**, **93**, **95**, **96**
Shaw, William Napier, 111
Shelter Island (N.Y.) conference, **215**
Shepard, Alan Bartlett, Jr., **339**
Sherwood (controlled fusion project), **208**
Shockley, William, 182, **281**, 283, **283**, **422**
Siegbahn, Kai, 284
Siegbahn, Karl Manne Georg, 281
Skylab, 230, **341**, **348**
Slipher, Vesto Melvin, 60, **101**
Small Magellanic Cloud, 59, **86**, **87**
societies, scientific, 259–260, **389**, **391–392**
Soddy, Frederick, 4, **10**, 14, 284
software, 196
solar radio observatory, **322**
Solvay, Ernest, **61**, **394**
Solvay conferences
    fifth, **61**
    first, **394**
Sontag, Henriette, **48**
Southworth, George, **278**
Soviet 6-meter reflecting telescope, 243, **381**, **382**
Soyuz (Soviet spacecraft), 331, **420**
spaceflight, 219–220, 229–230, 260, **327**
    exhibits, **417**, **418**, **420**
    experiments in space, **46**
space quantization, **58**
space station, 220, **332**, **341**, **420**
space suit, **260**
Space Telescope, 244, **365**
spectra
    expanding universe, **104**
    hydrogen, **28**
    quantum transitions, 33–34
    quasar, **377**
    solar, **113**, 230, **344**
    stellar, 71–72, **110**
    X-ray, **23**
    See also Zeeman effect
spectrographic instruments, rocket-borne, **344**
spectroheliograph, **113**
spectrophotometer, infrared, 158
spin, 34, 123

spiral nebulae. *See* galaxies
Sputnik, 220, 260, **321**, **329**, **418**
Stanford Linear Accelerator Center, 144, **219**
Stanley, Wendell Meredith, 285
Stark, Johannes, 281
stars, 71–72, **106–112**, **370**
star streaming, 71, **94**
Staudinger, Hermann, 158, 168, **234**, 285
Stein, William Howard, 286
Steinberger, Jack, 284
stereochemistry, 157
Stern, Otto, **58**, 282
Stommel, Henry, 92
stored-program concept, 196, **300**
Strassmann, Fritz, 124, **193**, **194**
stratosphere, **167**
strong force, 123
Strutt, John William, 84, 281
Sumner, James Batcheller, 285
sun, **113–117**, 230, **345–348**
superconductivity, 14, **30**, 158, **245**
superfluidity, 14, **32**
Supernova 1987A, **229**, **370**
Super Proton Synchrotron, **223**
Surtsey (volcanic island), **165**
Surveyor (U.S. moon probe), 220
Svedberg, Theodor, 284
Sverdrup, Harald, 92
synchrotron, 143
    electron, **211**
    proton, **216**, **223**, **227**, **228**
Synge, Richard Laurence Millington, 285
Szilard, Leo, 124, 133

tabulator, punched-card, 195, **290**
Tamm, Igor Evgenevich, **207**, 283
Taube, Henry, 286
Teflon, 168, **260**
telescopes, 49–50, **71–84**, 89, 243–244, **365**, **366**, **371–373**, **379–382**, **414**
television, **265**, **277**
Teller, Edward, 134
tensor algebra, **47**
Tesla, Nikola, 209
tests, analysis of results, 195
Texas Instruments, **285**
Tharp, Marie, 102
Theoretical Physics, Institute for (Copenhagen), **59**, **66**
Thermos flask, **29**

Thomsen, Julius, **19**
Thomson, George Paget, 13, 34, **62**, 282
Thomson, Joseph John, 4, **13**, **14**, 20, **25**, **272**, 281
Ting, Samuel Chao Chung, **221**, 283
TIROS (U.S. weather satellites), **178**
Tiselius, Arne Wilhelm Kaurin, 285
Titov, German Stepanovich, **330**
Todd, Alexander Robertus, 285
tokamak, **207**, **209**, **210**
Tombaugh, Clyde William, **121**
Tomonaga, Shinichiro, 283
Townes, Charles Hard, 283
transistor, 182, **281–283**, **308**
transmutation of elements, **11**
Trieste, 91, **135**
Trinity bomb test, **195**, **198**, **200**
Truman, Harry S., 134
Trumpler, Robert, 60, 72, **112**
Tsiolkovsky, Konstantin Eduardovich, 219, **325**, **327**
Tsukuba Science City, 272
tubes, electronic, 181–182, **266–268**, **272**, **274**, **276**, **277**
tube worms, **142**
tungsten, **242**
Tunguska meteorite, **118**, **119**
21-centimeter line, 210, **317**, **319**
Twin Tail radio galaxy, **383**

Uhlenbeck, George, 34, 123
ultraviolet light
  sun, **346**
  Venus, **350**
uncertainty principle, 34, **48**, **59**, **60**, **64**
unified field theory, 24, **47**
uniformitarianism, 83
UNIVAC, **305**
universe, 59–60, 244
uranium, 3, 4, **5**, 133
Urey, Harold Clayton, 14, **27**, **28**, 284
Ussher, James, 83

V-2 (rocket), 220, 229, 230, **328**
Valles Marineris, **354**, **356**
vanadium, **57**
Van Allen, James, 229–230, 260, **342**, **343**
Van Allen radiation belts, 260, **342**

Van de Graaff, Robert, 143
Van de Graaff generator, **408**
van de Hulst, Hendrik Christoffell, 210
van den Broek, Antonius, 4
van der Meer, Simon, 284
van der Waals, Johannes Diderik, 30, 281
Vanguard, 229
van't Hoff, Jacobus Henricus, 157, 284
Van Vleck, John Hasbrouck, 283
Veksler, Vladimir Iosifovich, 143
velocity-distance relation, **102**
Velox (photographic paper), 168, **257**
Venera (Soviet Venus probes), **349**, **352**
Venus, 210, **350–352**
Verein für Raumschiffahrt (Society for Space Travel), 219
Verschaffelt, J. E., **61**
Very Large Array, 210, **323**
Viking (U.S. space probes), **356**, **358**
Villard, Paul, 4
Vine, Frederick, 92, 102
Virgo galaxy cluster, **91**, **374**
Virtanen, Artturi Ilmari, 285
viscose, 168
Voigt, Woldemar, **389**
Von Braun, Wernher, 219, 220
Von Neumann, John, 112, **300**
Vostok (Soviet spacecraft), **330**, **406**
Voyager (U.S. space probes), 230, **359–364**

Wallach, Otto, 284
Walsh, Don, **135**
Walton, Ernest Thomas Sinton, 143, **188**, 282
Warner & Swasey, 272
water, **232**
waveguide, **278**
wave-particle duality, 13, 34, **48**, **52**, **62–64**
weak force, 123
weather. *See* meteorology
Wegener, Alfred, 101, **153**
Weinberg, Steven, 284
Weizsäcker, Carl Friedrich von, 72
Werner, Alfred, 284
Westerbork Synthesis Radio Telescope, **320**
Wheeler, John, **215**, **367**

Whipple, Fred, **357**
Whirlwind, 196, **302**, **304**
White Sands Missile Range, **344**
Wideröe, Rolf, 143
Wiechert, Emil, 102
Wieland, Heinrich Otto, 284
Wien, Wilhelm, 13, 281
Wigner, Eugene Paul, 283
Wilkinson, Geoffrey, 286
Williams, Frederic Callard, **301**
Willstätter, Richard Martin, 284
Wilson, Charles Thomson Rees, **12**, 13, **61**, 281
Wilson, John Tuzo, 102
Wilson, Kenneth Geddes, 284
Wilson, Robert, 210, 283, **387**
Wilson cloud chamber, **11**, **12**, 13, **183**
Windaus, Adolf Otto Reinhold, 284
Wislicenus, Johannes, 157
Wittig, George, 286
Wolf, Max, 271
women as "computers," **109**, 195
Woodward, Robert Burns, 158, **237**, 285
World's Columbian Exposition, **72**, **272**, **414**
World's fairs, 272, **413–416**, **418–420**
World War I, 167, **249**, 251
World War II, 209, 220, 260, **312**, **316**, **328**
W particle, 144, **224**, **225**
Wright, Wilbur and Orville, 219, **326**
Würzburg C radar, **316**
Würzburg Physical Institute, 3
Wynn-Williams, Charles Eryl, **15**

X-1 (experimental plane), **417**
xerography, **263**
X rays
  celestial, 244, **368**, **386**
  diffraction patterns, **51**, **52**
  discovery, **3**, **4**
  medical use, 181
  solar, **347**
  spectra, 23

Yang, Chen Ning, 283
Yerkes, Charles T., 49, **73**
Yerkes Observatory, 49, **72–74**
Yukawa, Hideki, 124, **187**, 282